SCIENTIFIC PAPERS *and* PRESENTATIONS

SCIENTIFIC PAPERS *and* PRESENTATIONS

MARTHA DAVIS
Department of Agronomy
College of Agricultural, Food, and Life Sciences
University of Arkansas
Fayetteville, Arkansas

ILLUSTRATIONS BY
GLORIA FRY

San Diego San Francisco New York Boston London Sydney Tokyo

This book is printed on acid-free paper. 

Academic Press
A Harcourt Science and Technology Company
525 B Street, Suite 1900, San Diego, California 92101-4495, USA
http://www.academicpress.com

Academic Press
Harcourt Place, 32 Jamestown Road, London NW1 7BY, UK
http://www.academicpress.com

Library of Congress Cataloging-in-Publication Data

Davis, Martha, date.
Scientific papers and presentations / by Martha Davis.
p. cm.
Includes bibliographical references and index.
ISBN 0-12-206370-8 (alk. paper)
1. Technical writing. 2. Communication of technical information.
I. Title.
T11.D324 1996
501'.4--dc20 96-28241
CIP

PRINTED IN THE UNITED STATES OF AMERICA
00 01 BB 9 8 7 6 5

To

Aaron Davis

Marion Davis

Jody Davis

Contents

6

THE GRADUATE THESIS

7

PUBLISHING IN SCIENTIFIC JOURNALS

8

STYLE AND ACCURACY IN THE FINAL DRAFT

9

REVIEWING AND REVISING

10

TITLES AND ABSTRACTS

11

PRESENTING DATA

12

ETHICAL AND LEGAL ISSUES

13

SCIENTIFIC PRESENTATIONS

14

COMMUNICATION WITHOUT WORDS

15

VISUAL AIDS TO COMMUNICATION

16

THE SLIDE PRESENTATION

17

POSTER PRESENTATIONS

18

GROUP COMMUNICATIONS

19

COMMUNICATING WITH OTHER AUDIENCES

Preface

Whether it is a chemical structure, the anatomy of a rose, or an odyssean siren hidden in the recesses of a DNA code, something stirs the curiosity and lures people into the realm of science. Most scientists are effective, intelligent, logically thinking individuals who are coordinated enough not to destroy the laboratory or the field plans and samples, but many become frustrated with communication. The siren did not tell them that many hours of their scientific days would be spent writing reports, preparing for presentations, serving on committees to solve problems, or telling the nonscientist about the value of the science. This book is an attempt to alleviate some of those frustrations with papers and presentations.

Because it is a single, relatively brief volume, I cannot hope to treat every kind of scientific communication in great detail. Other more definitive books concentrate on their respective subjects such as writing skills, journal article publication, writing proposals, group communications, public speaking, and all the other topics to which I have dedicated single chapters. My purpose here is to introduce fledgling scientists to most of the kinds of professional communication that will confront them during graduate studies and as career scientists. My objectives are (1) to answer the basic questions that might be asked about scientific communications and (2) to refer the scientist to more detailed sources of information.

To accomplish these objectives, the first part of the book proposes some practical ideas relative to preparing for, organizing, and producing a rough draft of any scientific paper or presentation. From these general concepts, the book moves to specific written forms that graduate students in science will likely encounter—the literature review, the research proposal, the graduate thesis, the journal article, and the practices and problems that accompany these forms. In scientific writing and speaking, it is important to understand publication styles, abstracts and titles, presenting data, reviewing and revising, and even ethics, copyrights, and patents.

Although a clear distinction between written and spoken forms of scientific communication cannot be drawn, I have concentrated in the latter part of the book on slide presentations, communication without words, effective visuals, poster presentations, and group communications. I include a chapter on communicating with nonscientists in both writing and speaking.

The appendices and the references are perhaps more important than they are in most books. The appendices provide additional information and examples on the topics discussed in the text. The references extend the views on communication beyond what I can include in this volume. I can give you an introduction to each topic, but you must go elsewhere for other details. At the end of each chapter are references cited in that chapter. Finally, following the appendices is an annotated bibliography of select works that I find most valuable. I am grateful for what I have discovered in all these sources and believe reference to them will be valuable to any scientist.

Everyone who begins a career in the sciences would do well to have had courses in technical and scientific writing, public speaking, group communications, graphic design, scientific presentations, journalism, leadership and interpersonal skills, professional ethics, audiovisual principles, rhetoric, and other subjects that develop the practical skills of communication. Because taking all these courses would accrue enough college credits for more than a degree in themselves, it is unreasonable to suppose that scientists will be trained in all these areas. They will have little time for practicing writing and speaking skills beyond the efforts required by their work. Similarly, the reading load for a given specialized area of science does not allow time to read all the books that have been written on the formats and skills used in scientific communication. This single handbook will often answer the questions that a graduate student or scientist would ask about scientific papers and presentations and will provide references that can lead to more comprehensive information.

I owe a debt of gratitude to all the graduate students in the sciences who have asked the questions I attempt to answer here. They have provided the motivation and stimulation that have prompted me to put my lecture notes in a form to provide assistance in communication for any graduate student, scientific neophyte, or even the seasoned scientist who may consult this book. I appreciate all that these students have taught me. A special thank you is in order for Terry Gentry, David Mersky, and Katie Teague, who have allowed me to use their work in my examples.

I also appreciate all the colleagues who have encouraged me in my teaching and writing, but especially helpful has been Duane Wolf. Without him I would never have pursued this or many other projects. His contributions to the subject matter and to my morale are immeasurable. Thanks, DCW.

Others have helped with the reviewing and revising, and I truly appreciate their time and attention. Thanks to Marion Davis, who read the whole thing, scrubbed out much of the wordiness, and complained very little the entire time. Thanks to Jody Davis, who read much of it too and was always there to buffer me from the word processor and to keep me motivated. Special thanks to Nora Ransom, who gave helpful suggestions on much of the text. Other reviewers on particular parts include Carole Lane, Sara Gragg, Rick Meyer, Lisa Wood, Lutishoor Salisbury, Bob Brady, Domenic Fuccillo, and Bill DeWeese. Suggestions from all of you have helped tremendously.

For the appendices, Justin Morris provided information for the abstract; Gail Vander Stoep allowed me to use her review of a manuscript; and the editorial, "Let There Be Stoning," is from Jay Lehr. I appreciate all of these contributions and the permission granted by Marilyn Hoch, Senior Editor of *Ground Water,* for the use of Lehr's editorial. I appreciate the permission from Walter de Gruyter & Co. for use of the excerpt from pages 69 to 73 of Eduard Imhof's book, *Cartographic Relief Presentations.* Thanks to E. M. Rutledge, M. A. Gross, and K. E. Earlywine as well as D. C. Wolf for the use of a photograph and the text of their poster.

For artistic contributions, I thank Gloria Fry, who has provided most of the illustrations, and also Steve Page for the computer drawing on posters. I appreciate the use of Martha Campbell's cartoon, which appeared originally in the *Phi Delta Kappan* (**73,** 130) and the one from A. (Toos) Grossman, which was published in *The Chronicle of Higher Education* on July 1, 1992.

Finally, a special thanks to Aaron Davis for his constant love and support.

MDavis, 1996

1
TO THE FLEDGLING SCIENTIST

"If it dies, it's biology, if it blows up, it's chemistry, if it doesn't work, it's physics."

JOHN WILKES
as quoted from graffiti on a bathroom wall.

Scientific communication is essential for helping us to use and take care of this earth. Researchers who discover the wonders of science must tell someone about their findings in clear, complete, and concise terms. To add to the pool of scientific knowledge, scientists must synthesize available information with what they discover. If a scientist garbles words or leaves out important points, messages become unclear, and the progress of science suffers.

No special talent is required nor is magic involved in clear communication. It's simply a skill developed for exchanging meanings with words and other symbols. Meanings associated with those symbols must be the same for both the sender and the receiver. But either the author or the audience can manipulate meanings, and being human, both probably will. Communication is the vehicle that carries progress, but it also carries disputes and disruption of progress. Generation gaps, wars, and prejudices result, at least in part, from something communicated. On the other hand, bridges across generation gaps, peace, and understandings are also results of communication. In scientific communication, be ever wary of the human elements, and communicate as concisely, conventionally, and clearly as you can.

Writing or speaking about scientific research is no more difficult than other things you do. It is rather like building a house. If you have the materials you need and the know-how to put them together, it's just a matter of hard work. The materials come from your own study and research. Any attempt to communicate in science is fruitless without valuable content. Once quality ideas and data are available, you put them together with the basic skills of scientific writing or speaking. The hard work is up to you.

Developing communication skills requires a combination of mental and physical activity. Like any such activity, it requires regular exercise or practice to move toward perfection. With swimming, you can't simply let someone tell you how, follow those instructions, and win a national title the first time you swim. The same is true with writing or speaking; only with continual practice can you develop and maintain the skills you need. Once you feel comfortable with those skills, you may even enjoy writing and speaking to an audience.

You have been in school for many years of your life; you know how to talk and write. You may or may not have had much of the needed practice in scientific writing, but you probably have had all the grammar and rhetoric courses you want. Don't disparage those courses. Basic instruction in language use is a good foundation for writing and speaking so long as you don't let that instruction inhibit your communication. Sloppy grammar, punctuation, and spelling can be very distracting to a scientific message. But this text does not presume to instruct you on points of grammar and basic composition, but rather on how to approach scientific communication and how to produce, review, evaluate, and revise papers or presentations. Those tasks can be easier as you define your purpose in communicating and develop guidelines that will work for you.

First of all, you need to come to terms with your purpose. Why are you writing or speaking about a certain subject? Obviously, several motivations stimulate your communication. Students often say their purpose is simply to fulfill an assignment. Maybe your reason for writing a thesis is to get a degree, or you are writing a journal article to get a promotion. Those are certainly good reasons to write. But surely purpose goes beyond those goals. A general purpose is the exchange of scientific knowledge; your specific purpose will depend on your subject and your audience. You may want scientists in plant breeding to know that cotton fiber initiation begins at anthesis. Once you define that purpose, you just need to develop ideas to answer questions that might be asked about that conclusion. The more specifically you define your purpose, the easier your task will be. When you have defined why you are communicating, your next job is how.

Any communication, and especially information exchange between scientists, is a matter of asking and answering questions. Answering the question before it is asked often averts many problems. In scientific communication, asking the questions is the foundation for discovery; providing answers to your colleagues and to future generations adds knowledge to knowledge and keeps scientific progress alive and well. From "How are you?" or "What's happening?" to "Have we discovered the final quark?" or "How important is preservation of the tailed toad?" the questions form the foundation for communication whether or not they are asked. If someone didn't wonder about answers, science would be in real trouble. As you consider a paper or a speech for your fellow scientists, decide what questions are in their minds and yours and which ones you can and should answer.

All forms of scientific communications have a great deal in common. Variations in content and organization are imposed by the questions from different audiences and the answers you give. An audience of sixth graders will not ask the same questions that scientists in your discipline will ask, but you can cover the same subject for both groups. In communicating about your work as a scientist, content and organization are clearly influenced by scientific methods of inquiry and reflect recognition of a problem, observation, formulation of a hypothesis, experimentation, collecting and analyzing data, and drawing conclusions. Notice that each of these steps poses a question that your research and then your communication seek to answer. What is the problem? What do you observe about it? What do you hypothesize? How do you experiment or explore for a solution? What data will you collect and how will you test it? What conclusions can be drawn? The content of your scientific paper will involve some or all of these questions no matter who the audience is.

Another major influence on organization and content is the use of conventional techniques in scientific communication. An audience can understand you if you use communication devices that they expect. For example, most organization, whether it is for journal articles, laboratory reports, or seminar presentations, uses the IMRAD format. The acronym IMRAD stands for Introduction, Methods, Results, And Discussion. These sections are the conventional, or the expected, order for scientific papers. A few journals alter this formula and present results and discussion before methods. They are simply answering the question "What did you find out?" before "How did you find a solution?" That organization does not negate the convention; it just asks the questions in a different order.

The IMRAD format is a common example of what is conventional or what the reader or listener expects. Much of the expected we do without realizing that we are following conventions. For example, in English the order we give to words within sentences is generally the subject first followed by the verb and then its object. Notice that the object in the second sentence of this paragraph does not follow that pattern. As a result, the sentence sounds a little strange, and it would perhaps be a better one if the order were conventional. Except for trying to call attention to a sentence, you should give the reader the expected.

A major purpose in this text is to outline what is conventional for the forms of scientific communications.

The rest is up to you. In addition to the questions from a given audience and the conventions that have evolved in language, your success in communications depends on knowing who that audience is, knowing your subject and purpose, and recognizing your own abilities and convictions. As you develop your career as a scientist, periodically remind yourself about the fundamentals of that science and about the fundamentals of successful communication (Fig. 1-1). Visualize your audience and consider your subject and your purpose for communicating. What questions will that audience ask and how can you best answer them? What media will best convey your message? Finally, every individual communicates differently; you need to think about yourself and your capabilities.

Think first about your **audience.** They are most important to the interpretation and understanding of your scientific message. However hard you try to send a clear message, the completed communication rests with them. You cannot control an audience entirely, but because you are initiating the commu-

"Class, who can tell me what Mr. Billingsley did wrong, in addition to majoring in this discipline."

FIGURE 1-1

Periodically remind yourself about the fundamentals of science and about successful communication. (Cartoon from Andrew Toos (July 1, 1992). *The Chronicle of Higher Education.* Used with author's permission.)

nication effort, you are responsible for presenting information that is easily interpreted and understood.

For most scientific papers and presentations, your audience will be scientists and often those especially interested in your subject. However, you will need to communicate also with other scientists and with lay audiences. Your grant proposal may be going to an agency that has no scientist on the staff or to a group of politicians or government officials. You may be trying to communicate with a publisher or an editor. You may need to transfer a new scientific and technical discovery to those who can make practical use of it but have little understanding of the science involved. Think in terms of how much experience and education the members of the audience have and what their motivation is for listening to you. Their attitudes and expertise can determine how you will present your subject.

Regardless of their prestige and education, members of the audience are human, and so are you. Human beings are rarely logical, fair, and unemotional. No matter how much you try to keep scientific communication strictly factual and objective, the human element is present. For example, if you are making a speech, the audience will notice your appearance and your voice before they ever hear a word you are saying. When readers look at a page, they notice appearance: the size of type, whether paragraphs are short or long, and whether there are headnotes or illustrations. People have certain expectations about how a speaker should dress and sound and how words on a page should look.

Once words are introduced, the reading audience or listeners have further expectations about meanings and patterns for those words. Most educated people expect standard English diction and constructions. If either is substandard or foreign to them, a break in communication results. Whether you are talking or writing, if you first give the receivers what they expect, or what they find familiar, they can feel comfortable. You can then lead them to your point even if it is unexpected or unfamiliar. It's not always the meaning of words that matters as much as it is the way we hang them together. No word can be fully defined except with the context in which it is sent and how the audience receives it. *The extent to which a word or idea reaches the audience with the same meaning it had when it left the sender constitutes clarity in communication.*

Other things to consider before you begin to write or speak are your **subject** and **purpose** in relationship to the audience. You have to be convinced that your subject is worthwhile, that the audience cares about it, and that what you are telling is accurate. For your own confidence you need a clear purpose. *Conviction* is a key word that joins with *concise* and *conventional* in producing clear communication. After you are convinced, the next job is to convince your audience. The same subject can be addressed to a group of scientific specialists or to third-grade science students, but not with the same words and techniques. The form your subject takes and the purpose you pursue will partly be determined by who the audience is. A main reason for always having someone review your work is to determine whether your subject has been adjusted to your audience.

With your audience, subject, and purpose in mind, remember that communication is essentially a question and answer format, but often the question is neither written nor voiced. Your communication will likely succeed if you are answering the same question that is being asked about your subject. That job may sound simple, but be judicious in deciding what questions to answer. The IMRAD organization for most scientific papers gives you the basic questions. You'll tell what you did in the Introduction, how you did it in the Methods section, and what you found out in the Results. It's all the little questions in between that can be frustrating. Who are you? Why did you do it that way? Is that like the result that Jones got? Can you give me an example? These questions are not always obvious ones that would be asked, but they require answers if minds in the audience are asking.

In addition to thinking about the audience and the subject and which questions they would ask about that subject, you also need to think about **yourself.** Attitude is as important in scientific communication as it is in all our activities and accomplishments. If you hate to write or speak, you won't do either well. If you love to communicate and find everything you write or talk about a true delight, you also won't communicate well. A scientific attitude of confidence but of careful self-scrutiny and a dash of humility can form the foundation you need for successful papers and presentations. Allow yourself to be creative, but keep an element of scientific control on your compositions.

You have to believe in what you are doing and what you are saying. Intelligence and personality go a long way, but when you get down to the nitty gritty of doing science and reporting it, you have to have conviction. This dedication can carry you far in science and always shows in your communication. Don't misjudge your audience. They can tell when you are bluffing, when you don't care, and when you don't believe in what you are doing or saying. You can learn rules for communication; you may even be good at play acting. But without some ability, integrity, and sincerity, your efforts in communication will fall far short of excellent.

Notice how all the elements in communication are interwoven. I can't talk about subject without discussing purpose and author and audience. Conviction and convention also depend on all those things. In fact, we understand any one of the major elements in communication only relative to all the others. There are no rules. There are questions to ask and answers to provide. An interaction of author and audience with subject and purpose through technique will produce communication. In this complex of influences, develop the skills to keep it as simple as possible.

This handbook surveys the forms of scientific communication that a graduate student or a scientist will most likely encounter, and it suggests approaches to the most common problems associated with developing necessary skills. The appendices that I have included provide examples of various kinds of papers or presentations and other ideas for polishing your communication skills. But all the instructions and good examples in the world cannot make you a good writer or speaker. A handbook may be able to tell you how to swim, but developing the skill is up to you. Above all, know how to keep your head above water or be able to come up for air when you need it. Your best source of information on those judgments is you. I hope this text is valuable in helping you to make critical distinctions between what is or is not effective in the scientific exchange of information.

2
BEFORE YOU BEGIN

"He has half the deed done, who has made a beginning."

HORACE

Before you begin to prepare the paper or presentation, you need to think about what kind of communication it is and who the audience is. That knowledge can lead you directly to sources of help with your communication effort. Also, you need to consider the tools available for producing such a paper or presentation. Word processors may not make the initial composition any easier, but they are a godsend for revisions. Before we get to the more specific efforts in writing or speaking, let's consider these issues.

KINDS OF SCIENTIFIC COMMUNICATIONS

The most common forms for scientific communications are reports, journal articles, proposals, theses, abstracts, speeches or slide presentations, poster presentations, and sometimes books, chapters, review papers, and group communications. **Report** is really a catchall term that includes everything from a laboratory account of a single experiment to progress reports and group reports on entire research programs. Your chief interest in graduate school may be the **thesis** or **dissertation** and the **proposal,** but as a scientist, you will also become concerned with producing **journal articles, abstracts, slide presentations,** and **posters.** Acquaint yourself with all these forms and produce them with precision, clarity, and always honesty. Your first attempts with scientific papers and presentations may benefit from some help. In addition to this handbook, become familiar with other sources that can save you time and improve your communication. I especially recommend those in the annotated bibliography at the end of this book.

SOURCES OF HELP

If you are just beginning graduate school, you should immediately acquire any information available through your graduate office or your department on writing theses, proposals, or other reports. Also, read *On Being a Scientist* (Committee on the Conduct of Science, 1989). Stock's (1985) guide to graduate studies can be very helpful, not only in talking about communication but also with all the other chores that a graduate program and research require. Smith (1984) has produced a similar book to consult as you get into graduate research. He has advice on communications as well as other problems having to do with research skills. Become familiar right away with the library services and computer data bases for literature searches in your discipline. Most libraries provide information about which data bases will be most valuable for you.

Read Day (1994). He writes the way he tells you to write—with brevity and clarity. O'Connor (1991) also has a great deal to offer about scientific writing. William Zinsser is not a scientist, but he's especially good relative to writing clearly and concisely. Read his *On Writing Well* (1988a) and maybe *Writing to Learn* (1988b). Others have helpful advice about scientific communications. Try Booth (1993) for practical instructions about both writing and speaking and Tichy (1988) for organization and other details on composition. Brun *et al.* (1984) have important things to say about speaking to a scientific audience, and Woolsey (1989) provides good information on poster presentations. If one author frustrates you, try another. You may find that Elbow (1981) or Ebel *et al.* (1987) present ideas that appeal to you more than my other choices do.

Know what the good reference sources contain so that you can use them as you need them. Your word processing program will probably have a spell check and grammar check. Use them, but they will not always answer your questions. Collect some references to keep handy. Begin with an unabridged dictionary and a good grammar handbook. The composition handbook you used as a freshman may be good. I recommend one like the *Little, Brown Handbook* (Fowler *et al.*, 1992). In addition, you will find a book on technical writing can be helpful for many of your professional communications. I like the one by Burnett (1994) called *Technical Communication*. I'm sure others are equally as good. *Scientific Style and Format* (CBE, 1994) contains detailed information on usage both in the language and in scientific conventions. Day's *Scientific*

English (1992) is a pleasant review of English usage for scientists. For manuscript preparation and documentation, perhaps no source is more detailed than *A Handbook for Scholars* by van Leunen (1992).

Special areas of science will have other handbooks peculiar to their disciplines, and some books, like that of Reynolds and Simmonds (1983) provide basic principles for presenting data. Before you begin to design your first graph or scientific illustration, check the Council of Biology Editors' standards in *Illustrating Science* (CBE, 1988). That book will answer far more questions than I can on graphs and maps and other illustrations. When you are ready to publish, check Day (1994) and O'Connor (1991) again.

Style, as presented in the composition handbooks, may not correspond with that in your scientific area. Learn about the style manuals or style sheets that you need to use. *Scientific Style and Format* (CBE, 1994) is a basis for style in all of the sciences, and it contains a list of style manuals for special disciplines. You may need to communicate with specific publishers for style sheets peculiar to their publications. Government documents, for example, have unique requirements. If you plan to publish in a journal, check the format and style of the publisher before you begin to write. When writing for journal publication, have handy any instructions from your publishing company and a copy or two of the journal to which you are submitting a manuscript. Often it's easier to see how an element of style is handled in the journal or book than to find a rule for it.

TOOLS FOR COMMUNICATION

You need to keep up to date with sources of information that can help you, and you also need to keep current with changes in the language as well as the media and the equipment available for retrieving information, composing your own paper or presentation, and disseminating your message to others. Language is your primary tool in communication, but other equipment helps you to transfer language from your mind to the minds in your audience.

English teachers have always had a problem with change. It is almost impossible to keep up with what is considered growth and development of language by some or desecration of the mother tongue by others. Be alert to all that is new, but I encourage you to err on the conservative side. As Alexander Pope recommended: "Be not the first by whom the new are try'd,/Nor yet the last to lay the old aside." Whether we like it or not, language is going to change from generation to generation, and as use of a term or construction becomes prevalent, little is gained by objecting to its acceptance.

Particularly important is knowledge about electronic communication and other technologies that can serve your papers and presentations. The computer age has brought with it new vocabulary. There are bits and bytes and DOS.

Acronyms become words; permuted words become words. Nouns become verbs. "Input" used to be just a noun; it's a verb now whether or not your dictionary says so. "On-line" is an adjective that describes a condition of the human being; "I'm on-line now" is almost as common as "I'm late." "Fax" was not long a noun before it also became a verb with a complete conjugation. It's the English teacher's nightmare. Nonetheless, Zinsser (1983) is right; we must accept word processing, electronic publishing, and the consequences of them.

The effects of this technology on professional communication are evident. We sing the praises of the Information Age. We can send and receive information at a phenomenal rate. Scientists are just a phone call away from each other and just a fax away from being able to share data. Literature searches are no longer limited by the time one has to plow through hard copy indexes or by the holdings of one library. The revising and editing of manuscripts no longer require scissors and paste.

We need to keep a realistic perspective relative to appropriate technology (Fig. 2-1). All the hardware and software in the world will not produce quality research nor write the paper for you nor make the successful presentation to an audience. We must have a human mind, a human hand, and sometimes a human face with the machine. Look at it from the same perspective our forebears must have had when the free-flowing pen and the pencil replaced the quill. Scribes must have been amazed at what the printing press could do for their work in the late 15th century. For them, moveable type would have been ingenious, and photocopy would be a miracle. We are fortunate to have made many advances in communication tools with binary systems, Boolean logic, and even the recent developments in fuzzy logic, but none of these tools alone will initiate the communication nor give to it an accurate interpretation. Don't lose sight of the importance of input and reception by human beings.

FIGURE 2-1
All the hardware and software in the world will not write the paper for you.

Take advantage of available technology. As a scientist you will be handicapped without access to a network, to e-mail, and to a fax machine. Your own personal computer can give you access to information retrieval systems through libraries and data banks. With telecommunication, you can carry on a meeting with more than one other scientist without leaving your office. You may even be able to activate a connection to your colleague simply by voicing a name. For teaching and group meetings, multimedia techniques now allow you to incorporate several audio and visual aids into one package.

Learn to compose at the computer. It erases easily and does not leave smudges. Learn to revise and edit from the screen. With data being easier to acquire and analyze, the dangers increase in trying to present too much in a short time, of being too wordy, or of putting too much information on a single visual. Make hard copy to proofread after you have proofread from the screen. Use the grammar and spell checks, but recognize their limitations. The spell check will not put the *y* on *the* to make *they,* nor will it retrieve a word that has been omitted. You may have the problem of the young scientist who presented his research in the poster format at a national meeting. He meant for his objective to begin with "To assess . . . " but it actually began with "To asses. . . ." He had spell checked his text; the audience at his poster smiled and were distracted from his scientific message. One person suggested changing the spelling of *to* to *two.*

In addition to composing your manuscript via word processing, publishing from camera-ready copy or from computer discs has several advantages. Traditionally, hard copy submitted to a publisher had to be reset into type for publication. This process provided opportunity for errors that cannot be made if camera-ready copy or material on disc is published without resetting. With the disc or camera-ready material, once the initial proofreading is done, the integrity of your copy should remain intact throughout the publishing process. The author can feel greater control over final forms of the publication. As technologies advance, publishers will be able to economically add new dimensions to your published work. For example, traditionally the cost of publishing color has been prohibitive. But color printers are now available, newspapers are printing color now; quick copy shops are offering that option. Color will probably soon appear regularly in graphs and photographs of scientific journals.

For your own use, desktop publishing equipment and software programs are now available that will produce text of high quality. Camera attachments or film recorders with computer units can photograph for slides or prints. Letter-quality and laser printers produce copy that is ideal for posters or slide copy. Some copiers can produce color illustrations for papers or posters directly from a disc or from a slide or photograph. With interactive media you can produce presentations that combine visuals, words, voices, and video action. Company representatives are always glad to discuss what their equipment will do, and if you attend meetings of your society, you will likely see demonstrations displayed. Personnel at local copy shops can explain what their equipment will produce. Find what will work best for your communication effort.

With all these advancements in technology, the tools for communication are decidedly improved, but the basic principles remain the same as they were when the scribe carefully copied the wisdom of Pythagoras with a quill to preserve and communicate it for new generations. Make full use of the keyboard, the mouse, and fiber optic networks, but don't simply default to what a program wants to do. Good software will allow you to make decisions yourself. Do so, and don't lose sight of what the Chinese and Greeks already knew when history was a child: *Simplicity and clarity are essential in conveying a message.*

Be alert for possible problems. Backup your input regularly on additional discs. Keep hard copy and don't throw away your pencil. Paper and pencil are not nearly so prone to technical disaster such as losses due to electrical surges or illnesses from viral infections. Not all equipment is without fault. When you attach people to imperfect equipment, problems can arise. The power of information exchange can be used in unethical ways. Copying software and invading privacy through stolen access codes certainly have occurred. Misinformation, faulty information, and misuse of information are dangers in electronic communication just as they are in books and other hard copy.

Check carefully when you are responsible for buying equipment and software for producing slides, photographs, or hard copy. Keep in mind criteria important to clear communication. Most important of all—don't let the technology dictate to you what constitutes good communication. We should not accept graphs that are too complex simply because the computer has the capability of producing them. We should not accept distortions simply because your computer program will not produce the expected dimensions or form. Manufacturers and programmers must continue to work toward providing capabilities that convey scientific communication in the most precise, understandable terms. The technology should in no way dilute clarity. Study the techniques for clear communication, and then make the equipment work for you.

Keeping up with what's going on in electronic communication is no easy task. As a scientist, you may need to rely on an electronic technician rather than be one, but find the programs that will serve you with informational retrieval, data processing, word processing, and dissemination of information. In this text I will deal with age-old principles of communication and hope that you can apply them to our changing world.

References

Booth, V. (1993). "Communicating in Science: Writing a Scientific Paper and Speaking at Scientific Meetings," 2nd ed. Cambridge Univ. Press, Cambridge, UK.

Brun, L. J., Bovard, R. W., Foss, J. E., and Stark, S. C. (1984). “Scientifically Speaking.” SRI, Inc., USA.

Burnett, R. E. (1994). “Technical Communication,” 3rd ed. Wadsworth, Belmont, CA.

Committee on the Conduct of Science (1989). “On Being a Scientist.” National Academy of Sciences, National Academy Press, Washington, DC.

Council of Biology Editors (CBE) (1988). “Illustrating Science: Standards for Publication.” CBE, Bethesda, MD.

Council of Biology Editors (CBE) (1994). “Scientific Style and Format: The CBE Manual for Authors, Editors, and Publishers,” 6th ed. Cambridge Univ. Press, Cambridge, UK.

Day, R. A. (1992). “Scientific English: A Guide for Scientists and other Professionals.” Oryx Press, Phoenix, AZ.

Day, R. A. (1994) “How to Write and Publish a Scientific Paper,” 4th ed. Oryx Press, Phoenix, AZ.

Ebel, H. F., Bliefert, C., and Russey, W. E. (1987). “The Art of Scientific Writing.” CH Publishers, New York.

Elbow, P. (1981). “Writing with Power: Techniques for Mastering the Writing Process.” Oxford Univ. Press, New York.

Fowler, H. R., Aaron, J. E., and Limburg, K. (1992). “The Little, Brown Handbook,” 5th ed. Harper Collins, New York.

O’Connor, M. (1991). “Writing Successfully in Science.” Harper Collins *Academic*, London.

Reynolds, L., and Simmonds, D. (1983). “Presentation of Data in Science.” Nijhoff, The Hague.

Smith, R. V. (1984). “Graduate Research: A Guide for Students in the Sciences.” ISI Press, Philadelphia.

Stock, M. (1985). “A Practical Guide to Graduate Research.” McGraw–Hill, New York.

Tichy, H. J. (1988). “Effective Writing for Engineers, Managers, Scientists,” 2nd ed. Wiley, New York.

van Leunen, M-C. (1992). “A Handbook for Scholars.” Oxford Univ. Press, New York.

Woolsey, J. D. (1989). Combating poster fatigue: How to use visual grammar and analysis to effect better visual communications.*Trends Neurosci.* **12,** 325–332.

Zinsser, W. (1983). “Writing with a Word Processor.” Harper & Row, New York.

Zinsser, W. (1988a). “On Writing Well,” 3rd ed. Harper & Row, New York.

Zinsser, W. (1988b). “Writing to Learn.” Harper & Row, New York.

3

ORGANIZING AND WRITING A ROUGH DRAFT

"Brilliance has an obligation not only to create but also to communicate."

J. R. PLATT

You can do a variety of things before you actually begin to organize and write a paper or put together a speech or a poster. A common choice is to procrastinate. You can even rationalize that the procrastination is leading to better communication. The paper is due soon and so you take a nap to get a fresh start on it. Maybe you will have a beer with a friend just to relax so that you will do a better job on the paper. You are even willing to wash windows or do laundry to distract yourself and thereby clear your mind for the important paper. The computer is right there, and so you play a computer game; you deserve the little break before the large task before you is begun. Be creative; you can surely think of many other diversions and rationalize your way into a most artistic, well-crafted procrastination. But at some point, you'll have to give in to the need to produce the paper. For the good student or scientist, self discipline will win despite the numerous skirmishes that procrastination wages in opposition.

Once that battle is won, consider what you may need for the paper or speech at hand. Determine what kind of paper you are to produce, isolate your purpose, and reflect on your subject and your audience. It may be helpful for you to study the examples provided in the appendices. Also, become conscious of the common weakness described in Appendix 1 so that you can be aware of them in your own writing. Then you will need to set your mind on the paper itself and perhaps go through a few prewriting exercises to condition you for producing a rough draft.

THINKING AND WRITING

You think well; you write well. I'm not sure that clear thinking depends on clear writing, but I am convinced that clear writing depends on clear thinking. And, like Zinsser (1988), I believe writing helps you to think and to learn. Thinking alone allows you to skip around, to move at random through a batch of ideas. The physical symbols you put on paper help to direct your thinking. Some people can visualize physical situations and can work out mental details for such projects as building a barn or repairing a motor. More often, the carpenter will sketch or look at a sketch of the barn, list materials and numbers of board feet needed, or even stand on the building site to visualize the structure. The mechanic will almost always want to look at and listen to the motor.

Words on paper have a similar symbolic influence on thinking, learning, and knowing. The physical words keep ideas in perspective and contribute to logical, orderly arrangements of thoughts. However, some people avoid this learning device because they find that moving thought to paper is difficult. Some find the writing itself relatively easy, but getting started is difficult. Multiple influences go into writing a scientific paper—the research, the data analysis, what others have written, your colleagues' opinions, your audience, your ability with words, the time you have for writing, how tired you are, what other thoughts are in your mind—the list becomes endless. Juggling your thoughts and having them all fall into place neatly on a page are difficult, but what can't be done in the initial writing can be done through revision. You must create a rough draft before you can revise.

Students often tell me that getting started is the hard part. Some outline; some don't. Some wait for time to pressure them; some write better without the pressure. Write the way you do it best. But if you are having trouble deciding what is best for you, try some of the following ideas.

PREWRITING EXERCISES

Think before You Write

Again, your first efforts should be the thinking about the audience and about the subject. Go through a series of questions and see that you are satisfied with your answers. What form will the writing take? How do I talk to this audience? What are they asking? What are my motivations? Do I believe in what I am saying? Do I have the materials I need including the literature, the data, the understanding, the reference books? Don't linger too long over these questions and their answers, but think long enough that you don't resort to the excuse of not being ready to write. When you have all the auxiliary matters out of the way, think for a bit about your subject, and then go on to another prewriting exercise or start writing the draft.

Talk before You Write

Prime yourself by trying to tell the material to a colleague, a spouse, a friend, or your pet. You can even talk to yourself via a tape recorder. The tape will hold your thoughts until you need them again. Talking out loud can cause you to recognize a logical progression of ideas or what is missing from your information. Involving your own ears and perhaps those of someone else can help you hear what details need to be included and emphasized and the order needed for clarity. With or without input from your listener, write immediately after talking while the subject is still up front in your mind.

Brainstorm, Freewrite, or Make a List on Paper

Just start putting words down on paper. This process can consist of a graphic presentation of ideas in boxes or circles joined by arrows, or it can be a running text of sentences as they come to mind. You may want to list the questions you think the audience will want answered. Try note cards; they are easy to shuffle and organize. Once you have a physical list of ideas in one form or another, you will be better able to organize the material. The point in any kind of brainstorm writing is to get words down on paper. You may throw them all away when you really get organized, but they are the stimulus and the material for your organization.

Outline

You may begin by outlining, or the outline may follow the brainstorming, freewriting, or talking. Some people can use either a rough outline or a perfected one to organize before they write any full sentences. Others may use outlining as an intermediate step to revise a first draft. Write first and then outline and rewrite if that technique works best for you. Once you have produced the outline, don't let it take over. It needs to be flexible so that you can reorganize if need be.

In making a list or a rough outline to work from, often it is best to leave the computer screen and take up your trusty pencil and a package of note cards. You can physically arrange and rearrange the note cards, place them all in front of you, and see the first and the last at the same time.

Write a Rough Abstract First

An abstract is really a kind of outlining. The abstract requires that you write a sentence of justification, a statement of objectives, a reference to methods used, a list of most important results, and any conclusion reached. A draft of an abstract will set forth the main points about which the paper must elaborate. Consequently, it can get you started in an organized manner.

Start in the Middle

Some people have trouble introducing a subject but know what they will say in the main body of the paper. An introduction could be the last part written. Don't waste time by trying to produce a perfect paper from beginning to end at the first sitting. Start in the middle and fill in around the edges as you can. The need for revision is a fact of successful writing. The important thing initially is to get something down on paper to revise.

Get Rid of Your Inhibitions

Loosen up. Most of us have inhibitions about writing the same as we do about speaking to an audience. The source of those inhibitions often has come from the very people who intended to teach you to write. Even on papers you wrote for English class, busy teachers probably commented little on the content. Maybe they scrawled an "excellent" or a "good" at the top of your paper. I was also familiar with "very poor." Despite having little to say about your ideas or content, those same teachers' red pens bled freely across the mechanical errors in grammar, punctuation, spelling. And they preached the gospel of the Rules (Fig. 3-1). We often reach graduate school still cowering before the mighty Rules. And, alas! The rules are flexible. I have become a heretic. Although I rely on grammar handbooks to show me what is conventional, I don't believe in allowing rules to distract from the content. The best set of rules I have seen proposed for scientific writing was devised by Nora Ransom at Kansas State University. The following are the only rules you will find in this book.

Ransom's Rules for Technical and Scientific Writing

1. If it can be interpreted in more than one way, it's wrong.
2. Know your audience; know your subject; know your purpose.
3. If you can't think of a reason to put a comma in, leave it out.
4. Keep your writing clear, concise, and correct.
5. If it works, do it.

FIGURE 3-1
We often reach graduate school still cowering before the rules.

Whatever technique you use to get started, get started. Don't let any of the monsters on the fringes of your thoughts deter you. Thoughts of an unreceptive audience, thoughts of other things you could be doing besides writing, hang-ups on mechanics of good usage that your English teacher has instilled in you—all such devils should be driven from your path. The only way to get something written is to write. Do it your way. In your revisions you must be sure your paper or presentation is well organized and that all the language monsters have been subdued so that you are communicating well with your audience.

ORGANIZATION AND DEVELOPMENT

During the thinking and prewriting processes, you may also organize your material within a format. Organization may, however, be a chore that follows after prewriting exercises or even after a freely written rough draft. Organization produces a unified package that makes both sending and receiving information easier. In addition to the arrangement of content, you use the tools of language—words, sentences, and paragraphs as well as symbolic signposts like transitions or headings and subheadings—to direct the reader or listener through the organization and development of your subject.

Organization and development are concepts that cannot be separated. It may be that a second point in your organization will not be understood until the first point is developed. To keep the progression of ideas logical, keep the audience in mind at all times. Think of someone listening or watching you write who is asking questions repeatedly: "What did you do?" "How did you do it?" "What do you mean?" "What caused that?" "Can you give me an example?" Answers to the questions constitute development of ideas, of sentences, and of paragraphs. This inquisitive little voice over your shoulder can help you keep things in order and tell you how much development you need. It will remind you when you need to explain further or when you are proceeding too fast without making clear the background, the definitions, and the causes and effects.

In writing or reading a paper, we move from word to word, from sentence to sentence, and so on, building larger and larger units of material until the full paper is put together or understood. In organizing a paper, we use the reverse procedure. We first consider the largest units. To an extent, the kind of scientific paper will dictate the largest units. A journal article, for example, will almost invariably follow the IMRAD format. Proposals may dictate different framework, possibly Introduction, Justification, Background or Literature Review, Methods, and Conclusions. A thesis may require yet another primary organization as will a review article. The large sections of your paper may be more lucid if based on chronological order or a step-by-step process; they may be based on spatial arrangements or geographical order; or they may be divided by meanings (two or three comparisons or contrasts, causes and effects, or pros and cons).

When the framework is in place, we must consider each smaller and smaller organizational unit. At some point you can quit organizing and start writing or talking. That point depends on your personality, your expertise, and maybe the way your brain works. Some people start writing as soon as main headings are established; some need details down to the developmental topic for each paragraph and even subtopics within the paragraph. It all sounds very simple, but the important point is to be sure that you extend your organizing to the smallest unit you need before you attempt to move in the other direction and write the paper. Notice again that, as you **organize,** you proceed from the large to the small:

Sections
Subsections—Subsections
Subsubsection—Subsubsection—Subsubsection
Paragraphs—Paragraphs—Paragraphs—Paragraphs
Sentences—Sentences—Sentences—Sentences—Sentences
Words—Words—Words—Words—Words—Words—Words—Words

And when you **write,** you go from the smallest to the largest unit:

Words—Words—Words—Words—Words—Words—Words—Words
Sentences—Sentences—Sentences—Sentences—Sentences
Paragraphs—Paragraphs—Paragraphs—Paragraphs
Subsubsection—Subsubsection—Subsubsection
Subsections—Subsections
Sections

Main sections may be labeled with primary headings, subsections with secondary heads, and subsubsections with tertiary heads. You'll probably not need more specific labeling. Paragraph indentions indicate new development of an idea just as capitals and periods tell us where sentences begin and end.

You have likely developed your own methods for outlining or organizing. If you haven't or if your technique doesn't work well, read Tichy (1988) or try the following for arranging your content. First, simply write down the big headings—Introduction, Results, Justification—and then start making notes or lists under each heading. You may immediately see an outline emerging. You may, however, need to arrange your list further by subordinating some ideas to others, putting materials into categories, and giving an order to what needs to be told first, second, third, etc. Now an outline is emerging. If you are a tidy person, you may want a formal outline. An English handbook can provide guidelines for outlining.

CONTENT

Conventional patterns of content development lie inside the large sections of your paper. Although not all papers follow the IMRAD formula, I'll use it to illustrate typical points in development because it is common for scientific communication. For more specific treatment of content for these organizational sections, refer to Day (1994).

The **Introduction** should serve three purposes: (1) to call attention to the specific subject or hypothesis that you are to discuss, (2) to provide background and justify a study relative to its importance and the results of other studies, and (3) to list the objectives of your research project or to give the audience information on what you plan to accomplish in your paper.

The **Methods** section will be a recipe revealing how you acquired your data. Organization here is usually easy with step-by-step processes kept in the same order as the objectives were listed in the introduction. You can preface this procedure with a listing of materials used, conditions present, or design of the project. In addition to needing details in the materials and methods used, the audience will also be asking two major questions: Is this researcher's work credible, and can I use the same methods? To answer these questions, you must provide complete information on ingredients, actions, conditions, experimental design, replications, repetitions, and statistical analysis. Ask yourself: Could another scientist follow my words in this part of my paper and perform the same experiment with the same results? A well-written methods section will support a positive answer.

Results should be organized so that the reader moves from one display of data to another with logical development showing how your findings satisfy your objectives. Results may be presented in the same order as the objectives or the experimental procedures. Data are often presented in tables or figures, and the text will simply serve to tie the data to your objectives or to call attention to main points.

In addition to the valid experimental design that you present in your methods, your results and your **Discussion** will establish your credibility. Discussion is sometimes interwoven with results, or it will be in a separate section. The commentary will provide an interpretation of the results and show relationships with other research. Summarizing statements will tie together outcomes as depicted by the various data sets.

Discussion should present the overall significance of your work and help direct the thinking of your audience, but once you have made a statement on what your data mean, don't go too far afield with speculation. Show how your results fit into or compare with those from similar studies, and leave to the minds of your readers most of the speculation. Think of your paper as a piece of art placed in front of viewers. Be sure your objectives, methods, and results are clear, and then just let them read their own interpretations into it. On the other hand, don't hesitate to suggest a direction you want the receiver's thoughts to take; just don't overdo your own speculation.

Beyond IMRAD, be sure to make concluding statements at the end of your discussion or in a section called **Conclusions.** Enumerate these conclusions succinctly; they may be the points that stay longest in the reader's or listener's mind.

ORGANIZATION AND YOUR POINT OF EMPHASIS

Organization may fall apart unless it is tied together with a thesis. *Thesis,* in this instance, means the theme, the motif, the focus, the overriding topic around which everything in your paper should revolve. Again, for communication, art and science follow the same principle. A good musical composition or piece of literature carries an overriding motif and lets us hear it or consider it from several vantage points. The question that the scientific audience asks you is "So, what's your point?" Over and over you need to reiterate that point or thesis.

If we employ physical imagery to interpret this concept, we can think in terms of the blocks of information, the various parts of IMRAD for instance, that make up a composition. A common thread must be woven through these blocks to hold them together. Most scientific papers emphasize one or two main points. The hypothesis is one version of this point and the objectives are another. Methods provide another view, and results concentrate on what you found out about that point. Any discussion and conclusions must strictly adhere to the point. With concentrated focus on your thesis, the organization is not likely to stray.

TRANSITIONS

When we talk about organization, we are talking about a road map that will carry you and your audience through the various developmental blocks or units of materials that make up your paper. We need mortar or bridges to join the blocks, and we need signposts to direct the reader along the road. If you think of the theme as a thread, these connections are loops in the thread that sew one section or idea to the next. Those language devices that hold your paper together are **transitions.** They help to unify and move the paper through a series of ideas to a conclusion.

Transitions serve both to join parts of a subject together and to convey meaning. Like the Roman god Janus, transitions look in both directions at the same time, but they signal movement to the next idea in your paper. They may be conjunctions or prepositions that hold parts of a sentence together with distinctive meanings. For example, the short words *and* and *but* are transitions that carry opposite meanings. The same is true of *to* and *from.* Keep in mind the functions of transition as joining two parts together and carrying the message forward with meaning.

Beyond simple words as transitions, you can use phrases such as "on the other hand" or "in the same way." You can use full sentences: "Chlorine reduced the bacterial populations in the first experiment. In the second, we substituted fluorine." The full sentences are transitional, and they contain the transitional words *first* and *second.* You can also use full paragraphs as transitions between two more complex points. Within those paragraphs, you will likely find several smaller transitional elements.

Another transitional device is repetition. You can repeat the same word or phrase, or you can repeat the same idea in different words. This kind of transition is important to carrying a major point or motif throughout a paper, referring to it again and again to keep the reader focused on that point. Repetition in styles used for headings, in sentence structure, in symbols, or in colors and designs can serve as unifying elements and as transitions. Good transitions make a paper or a speech flow smoothly and keep the communication from being jerky or jumping from one point to the next. For a more detailed but brief discussion on transitional devices, consult Marshek (1982).

Whatever connections you use or however you visualize your unified paper, the overriding requirement for good organization is clear thinking with use of convention and logic. In other words, if you do what is expected in a reasonable manner, your writing will be most clearly understood by other reasonable people who speak your language with the same conventions. Whether the break in convention is a misspelled word, an awkward transition, or chaotic organization, it will distract from communication.

WRITING THE ROUGH DRAFT

Once you have corralled your thoughts with or without prewriting exercises and considered your organization, you are ready to write a rough draft of a short paper or a section of a longer work. At this point, let's assume that you have focused on one point of emphasis, you have thought through what you want the audience to know when you have finished, and you have considered what their questions are. You have gone through some prewriting exercises and have produced an outline or other organized pattern for your work or have made a list and then organized your ideas so that your audience moves with you from one point to another. With all this preliminary work done, begin to write immediately. Any delay at this point is distracting.

Keep in mind that your first production is a rough draft and will require revision. Perhaps the only rough draft ever set in stone was Moses' Ten Commandments. That document was divinely inspired; your words are not sacred. When you have written a rough draft, be willing to tear your words apart, throw them away, insert new ones, and rearrange them until they say what you want them to say.

Language is really not so haphazard as it sometimes appears. Lack of clarity results only when a speaker or listener (writer or reader) deviates from logical organization, from standard meanings of words, or from expected form or word order. Think clearly before you speak or write, and the audience should receive your message without misunderstanding.

Even after careful organization, some people have problems getting started. There are not hundreds of ways to approach writing. Don't waste time wondering how to start. In creative writing, approaches such as stream of consciousness or story telling can add artistic or philosophic dimension to writing, but for scientific writing, you will seldom need to know how to handle more than the following four approaches: **define, compare-contrast, enumerate, and give cause–effect.** Choose one of the four and start to write immediately. Don't ponder. Don't use up time at this point. Pick one (or two). The others are all in the back of your mind and can be summoned at a moment's notice.

Compare–Contrast
Enumeration
Definition
Cause–Effect

Numerous key words used in asking and answering the questions will control the development in your paper or presentation. You may develop ideas with evidence, example, detail, points, methods, classification, results, summary, reasons, alternatives, possibility, etc. All these words are related to the four approaches. Evidence may be cause and effect or a definition or an enumeration of points or a comparison or contrast. The questions from the audience may take many forms, but the techniques involved in the reply are limited. Your answer will characteristically begin with a general assertion followed by details to explain and clarify that assertion.

The audience and their questions will often tell you how to start or, at least, hint at an approach. "What is a quark?" (definition). "What caused that result?" (cause–effect). If you must choose the approach, as in Example 4 below, you usually know which one you can handle best or which will best serve the audience.

EXAMPLES

1. Question: What are your methods for ____________________?
Answer: Enumerate 1, 2, 3, or 4 methods; then define, probably by enumerating steps in a process, showing cause and effect, or comparing and contrasting.

2. Question: What are the differences in ____________________ and ____________________?
Answer: Define terms and then contrast differences probably through enumeration.

3. Question: What is ____________________ and how does it affect ____________________?
Answer: Definition and cause–effect.

4. Question: Discuss reasons for continuous flood in rice.
Answer: Here you have none of the four approaches, but you choose one or two and begin immediately. Probably here you need to enumerate cause and effect: (1. weed control; 2. less loss of nitrogen; 3. greater yields) or compare–contrast to intermittent flooding (enumerate again).

A common way to begin an essay is with definition. A safe way (i.e., it will always work) to begin a definition is to put the term into a general classification of related ideas and then point out its distinctive qualities.

Formula: **(term)** is **(class)** which **(distinction)**

Example: Factorial anova [**term**] is an analytical technique [**class**] in which two or more variables and their interrelationships can be identified. . . . [**distinctions**] (Continue to enumerate other distinguishing characteristics.)

Comparison and contrast or cause and effect almost always involve definition and enumeration. Both approaches lead to examples and **tangible details.** A person can get lost in technical terms if tangible images are not occasionally formed in the mind. A tangible image is one that relies on the senses (or a past sensory experience) for understanding. We see, taste, smell, hear, and touch tangible examples. Words like lemon, freight train, rose, fuzzy, or square throw images into our consciousness to increase our understanding of an idea or concept. As you experience scientific and intellectual concepts, your senses are still at work. Terms like *hydrogen sulfide* or *staphylococcus* are understood in part through your senses of smell or sight once you have encountered the real substance. Notice the difference as we add tangible details in the following:

General:	Factorial anova compares relationships of two or more factors.
More specific:	Factorial anova was used to determine interrelationships of two treatments on three varieties at two locations.
Most specific:	Factorial anova was used to determine interrelationships between foliar applications of nitrogen and boron on 'Forrest,' 'Davis,' and 'Clark' soybeans in field plots at Fayetteville and Marianna.

As your communication becomes more specific, you alert minds in the audience to specific images that define and explain your message. No matter how theoretical your scientific communication must be, keep as close to tangible details as possible. If the senses can't be involved to make the details tangible, at least be as specific as possible with intellectual concepts.

A great deal of time is wasted in struggling with the first rough draft. Don't worry with perfection at this point; you simply need a draft to begin working with. If necessary, put a time limit on yourself, ask yourself the appropriate questions, and then write.

As you gain experience in writing, you will use the basic approaches to organization and development of ideas without realizing that you are doing so, just as you use periods at the ends of sentences and start the next with a capital letter without having to remind yourself each time. But when you are having trouble getting started or are revising a paper, it's good to think about the possibilities.

Checklist on how to organize and write a rough draft

1. Determine what questions you are answering for your audience and how specifically each question directs your approach.
2. List ideas that will convey the answers.
3. Arrange the ideas in a logical sequence.
4. Using your own judgment, choose one of the four approaches (define, compare–contrast, enumerate, give cause and effect) and write immediately.
5. Recognize the need to revise.

References

Day, R. A. (1994). "How to Write and Publish a Scientific Paper," 4th ed. Oryx Press, Phoenix, AZ.

Marshek, K. M. (1982). Transitional devices for the writer. *In* "A Guide for Writing Better Technical Papers" (C. Harkins and D.L. Plung, Eds.), pp. 107–109. IEEE Press, New York.

Tichy, H. J. (1988). "Effective Writing for Engineers, Managers, Scientists," 2nd ed. Wiley, New York.

Zinsser, W. (1988). "Writing to Learn." Harper & Row, New York.

4

SEARCHING AND REVIEWING SCIENTIFIC LITERATURE

"Do not condemn the opinion of another because it differs from your own. You both may be wrong."

DANDEMIS

PLANNING THE LITERATURE SEARCH

Any scientist needs to be familiar with the work that has been done in his or her area of research and with what others are doing concurrently. No scientist, however, can read all the books, articles, and other printed matter related to a given area and perhaps not even all that is concerned with his or her research. Therefore, we need to be selective and to find the most pertinent literature on a subject as efficiently as possible.

A deliberate, well-organized approach to a literature search can save you a great deal of time and energy. Searching the literature is a continual process, but first you will need to **explore ideas** related to your subject, then **conduct a specific search** in relation to your research objectives, and certainly **keep current** with what is published while you are pursuing your objectives.

An exploratory search can be conducted before a clear hypothesis and clear objectives have been established. During this exploration, you will look for gaps in research on a subject of interest and decide what specific experimentation to propose and pursue. In exploring the literature and in the more specific search that follows, you need to look for any duplication of your efforts. If someone has already done work similar to what you propose to do, you may need to change your plans to avoid reinventing the wheel.

The specific search is essential for providing background to your own work and for revealing where and how your research fits into the scientific pool of information available. It will begin with a survey of information associated with your specific objectives. As your research progresses, you will encounter even more specific needs for information. For example, you may be using a method and need to know other uses that have been made of that method or what new equipment is available to employ with your procedures. In such situations, your search becomes limited to a single point in your research.

Finally, when you are communicating your results, keep current with ongoing research. This current search can be maintained through habitual reading, a periodic computer search through current awareness sources, and interaction with your professional peers.

In designing an overall plan for a search, consider the following principles.

Visit the Library

This admonition may sound facetious, but I'm quite serious. Your personal computer may be connected to a network whereby you can do most of your searching through your own office, but you will do well to visit the library first. Browse around for 30 minutes, and check out the systems for finding and acquiring materials. Collect any flyers or brochures with information about the library and how it serves you. Determine where special materials such as government documents are filed and how to gain access to materials housed in other libraries. Interlibrary loans may be free, or a fee may be charged for acquisition of materials from other libraries. You may wish to browse through your library's collections or go on-line to browse at other libraries. A general familiarity with the library can be a real asset as you search the literature.

Allow Time

Often the scientist is so interested in what is going on in the research itself that the literature search is neglected. You can actually save time in some instances by consulting the literature to discover the latest data compiled on a subject or a new technique for accomplishing your goals. More important than saving time is your responsibility for knowing what has been and is being done in your area. This knowledge is essential to scientific communication whether you are writing a proposal for new research, reporting results of a study in a poster, making a slide presentation, or writing the journal manuscript about completed research. Give yourself time to explore, find, and read about your subject and perhaps time to write a review of the literature.

Isolate Your Objectives

You can waste time when you spend too much of it exploring before you establish a specific search or if you repeatedly get sidetracked on interesting subjects that are related but not directly associated with your work. Once you have declared a hypothesis and specific objectives, stick to them in your search. At least be sure that any interesting sidelights are worthy of the time you give to them.

Document Carefully

Make detailed notes on important works that you find. For a future bibliography you may need authors' full names, full titles, journal titles, volume and page numbers, publishers, dates of publication, and any other information that is relevant to the style of the publisher to whom you submit a manuscript. Any photocopies you collect should have all this documentation on them. Photocopy is a great convenience, but overuse of the practice can be detrimental to writing a review of the literature. Voluminous piles of copied materials can be just that—unorganized piles. Develop some system for filing photocopies so that important information does not get lost again after it has been found. Note cards are handy for compiling information from the articles and for making the first draft of a bibliography. They can be easily alphabetized and amended. Or you might want to invest in a software program for compiling and filing your references.

Be Selective

Not everything written about your subject is relevant to your work. Some publications can be substandard, and reference to them can jeopardize your own credibility. Leaving out important works can be as bad as including questionable work. Discriminate relative to source, author, date, and relevance to your work. Other discriminating scientists are familiar with the literature and with the scientists who are working in a given area. They are not impressed by work padded with irrelevant literature nor with that which fails to give reference and credit to the pertinent studies.

Verify

The most important principle that you can follow in searching and using the literature is **accuracy.** Careless use of ideas that allows ambiguous interpretation can injure your reputation. Careless documentation that leads fellow scientists to waste time pursuing a faulty reference should lead to your obliteration from the scientific community. Whatever style you use, be sure to include not only accurate but also enough information for a reader to find the source to which you refer. An accurate and complete reference is as essential to your writing as an accurate measure is to your laboratory research. To load your bibliography with what you think may be proper spellings of names or proper volume and page numbers is as dishonest as loading your scientific data. Verify all ideas and documentation.

Be Willing to Quit

Almost any library search can be extended indefinitely. Don't leave a search before the job is done, but don't continue to follow blind leads that give you fragments of remotely related material. At some point you must be willing to discontinue the search temporarily and apply what information you have gained to your research and to your writing. You can and should always go back for more and for an update.

Create Something Useful

From your notes and documentation, put together a complete bibliography. Sort out articles and books directly related to your own study and write a literature review on the subject. This writing can be important to a proposal, a thesis, and a subsequent journal article.

Verify

This principle is so important that it needs to be considered repeatedly as you collect the literature and when you use it. Check your bibliography and literature review against the original sources. Be sure there are no errors. Verify every time you revise or create new copy.

Keep Up to Date

We are all familiar with the college professor who is still lecturing from notes he or she made as a graduate student. Don't fall into this trap. Many thousands of books and articles are published every year on scientific subjects. Some of those relate to your work. Find them.

FINDING THE LITERATURE

In searching the literature, as with any job you do, you need to be familiar with the terminology and sources of information. Research findings and other scientific information can be published in books, monographs, proceedings, journals, dissertations, patents, standards, governmental bulletins or reports, and a variety of other forms. Some may be stored on microfiche or video- or audiotapes; more and more we will find information in electronic forms. Methods for searching each of these sources may differ greatly. One may require a search through a card catalog, another an on-line computer search, and still another a special index.

For a thorough search of the literature, you will probably do both manual and computer searches. Before you do either, you may need to decide where to look—what data bases to call on and which indices to consult to find the specific literature you need. Various guides provide information on appropriate indices and abstracting services as well as the primary literature. The library will have an information or reference section that will provide help. Library science is a relatively complex discipline, but librarians with expertise in the field are usually eager to assist you with a search.

Most libraries use similar filing systems, but each has its own unique characteristics. A journal you find at one library may not be in another. One library may utilize access to computer services from a different vendor than another uses. Most of them can now provide access to machine-readable information either locally on CD-ROM or via remote access. They have annotated listings of the data bases they access that explain what information each provides. Librarians can recommend indices for manual and computer searches. They can assist by showing you how to formulate search strategies, how to identify what is available and how to access it, and how to choose relevant key words and how to use Boolean operators. They can also provide you with thesauri that will help you isolate the appropriate key words for the search. Many libraries will provide a specialist in your subject area. Find out what and who are available at your library and use these services.

Computer data bases are certainly a godsend for any literature search. Recognize all their advantages. They can be time efficient and generally access a wide scope of the literature within a discipline. By using free-text searching and controlled vocabulary key words, you can execute a search strategy. Acquiring the literature itself to read and incorporate in a review of your subject is also made easier with machine-readable copy. Full-text access is available through many data bases, and new possibilities are becoming realities every day. With most networks, you can search from a personal computer through the holdings at your campus library and even at other libraries associated with yours. For computer searches, as for other electronic communications, be sure to keep up with what is available. What I say here may be outdated next year.

Although you can acquire voluminous information in minutes via computer, you should also recognize the limitations of computer searches and compensate with a search outside their boundaries. Without proper planning, you may acquire voluminous citations that are mostly irrelevant to your specific interest, and it will take time to sift the chaff from the grain. In addition, most data bases were not established until the late 1960s or the 1970s. The extent to which they have retrospectively converted information before these dates to a machine-readable format may be limited or nonexistent. Earlier works are still listed in printed abstracts and indexes. Consider the appropriate time period to search relative to your subject. You want to be on the forefront of knowledge in your area, but that condition may require that you look into important work that was done before the 1960s or information in special publications not filed with a computer data base.

With these cautions, consider an approach for your own search. Your specific search begins after you have explored a library and have established a hypothesis and specific objectives for your research. You or one of your advisors or colleagues probably knows of someone who has written about your subject in the recent past. A convenient way to start your search is to find that book or article that has been published recently and, not only read it, but also use any list of references it gives to find other articles. Use these other articles in the same way to create the **snowball** (Smith *et al.*, 1980). Each source yields another group of references, and in many instances you can accumulate a great deal of material on a subject simply with a snowball search.

This approach is limited in that you are always going back in time and the subject matter may get farther and farther from your own immediate question. This limitation may actually be an advantage when a manual snowball technique is used with a computer search through data bases that have not filed information previous to the 1960s or 1970s. Another way to compensate for the lack of recent literature in the snowball search is to seek names and key words through the *Science Citation Index.*

A second possible disadvantage to the snowball search is that sometimes scientists limit their references to persons closely associated with their own work or in a given society, and you may miss relevant publications in other countries, in other languages, or in sources outside the realm of those being cited by the group in your snowball.

After you have done a snowball search, you'll begin to feel comfortable with the terminology and names of the researchers who work in your area. Then you may do a further cursory computer search to check authors' names and key words in recent indices. When you are familiar with the terminology used with your subject so that you can isolate a limited number of key words and synonyms and plot them into a logical plan, you are ready for a full computer search. If you have had little or no experience with computer searches, a subject librarian can save you much time and frustration not only in the search itself but also in planning for it.

When you have sorted through the relevant and irrelevant information from your computer search, as well as the initial manual searches, you should have acquired a sizeable bibliography. However, if all your searching has not covered the full scope of time and content that you need, your next job is to hand search for materials outside those available through the computer. Your initial searches may have missed isolated publications important to your research, such as work in progress but not yet published or information published in a book or pamphlet not covered by the data bases or indices. You may be able to locate such references in a manual or computerized catalog. Current awareness sources such as SciSearch and Current Contents Search, CARL Uncover, ArticleFirst, ContentsFirst, and others will give you relatively recent information.

SELECTING THE LITERATURE

You will want to read completely any article that is closely associated with your study. However, you will retrieve numerous references that are not so important to you, but you won't know that they are not until you scrutinize them. Develop a method for saving time and finding out whether you should read the entire article. If the title appears pertinent, read the abstract. If that still indicates information that may be helpful to you, read the conclusions and look at relevant data in any tables and figures or at a method that may be important to your research. By the time you've done those things, you should know whether to read the entire article. Using this technique to screen articles can save you some time. But when you decide to cite a reference, you need to be thoroughly familiar with all that the authors are saying. Misquoting or taking a finding out of context can constitute an inexcusable inaccuracy.

So much literature is available on a given subject that it is often difficult to be selective. Base your selections on the credibility of the work and its relevance to your own. The literature may serve as a historical background to establish the position of your research in a larger framework. A reference may be selected as relevant if it teaches you a new method or gives you new ideas to pursue in accomplishing your own objectives. It may illustrate or justify a specific point you make in your work, or it may support a result you find, a method you use, or a conclusion you reach. Don't avoid citations that disagree with results from your own research. Determine why other researchers got different results and report the differences. A comprehensive discussion of the literature will increase your own credibility.

Select and collect all the literature relevant to your study. Judge credibility of a publication on the bases of who wrote the work, whether the source (journal or publisher) is reliable, and whether the methods and results reported are scientifically sound and well tested. If an article is written by a reputable scientist and published by a well-known publisher or a scientific society that follows a careful review process, that work is usually dependable. But check the methods and analyses with close scrutiny before you accept the information completely.

Most reputable scientific journals are published by professional societies or professional publishers. Scientific research done by government employees with the U.S. Department of Agriculture, the Department of Defense, or other agencies may be published as government documents. Distinguish between materials published even by these reputable sources. Societies may publish in lay magazines information that does not contain the complete data you will need to support your own contentions. They may also publish quick communications, letters, or proceedings that are clearly designed to disseminate information before research is completed and ready for journal publication. Commercial publishers are sometimes as thorough as professional societies in the review and editing processes necessary to ensure credibility, and they too will likely publish inconclusive but possible scientific breakthroughs in periodicals for the lay public. Distinctions in primary publication of research findings and speculations or interesting applications for the research are usually obvious. As a writer, you should be discreet in your use of these publications.

Some industries, companies, or corporations publish trade journals that can contain valuable information especially concerning what is available to research and development. One must accept in these journals a bias toward the company publishing the work without discrediting either that company or its competitors. Similarly, corporate institutions may publish information for their employees to help them keep up with research and development. Such corporations or companies may also publish educational materials for the public. All these publications provide important communication to their audiences, but the scientific writer must use clear discretion in making reference to them. Your own credibility is at stake.

Besides publications on work completed, know what is being researched on your subject. As you become involved with a profession, you will exchange information with your peers at professional meetings or by phone, with electronic mail, or through fax exchanges. You may sometimes use such personal communication as a reference in your own work. A bona fide, refereed publication is more likely to command the respect of your audience than a personal communication. But verified communication from an expert on a subject can be valuable support to your paper. We can expect that the results from quality research should be and usually are published. If, however, the information is an isolated finding or a very recent disclosure, it may be important but not yet published. Personal communications can also supply you with information on papers in review or in press or with research projects under way. A fellow researcher can refer you to literature you have overlooked or can help you perfect a method you are using. In association with your peers, you soon learn who is respected, and you aim for the same reputation. Your professional ethics must guide you in your openness as well as your confidentiality with your peers. Careful documentation of your sources is always vital.

THE LITERATURE REVIEW

Before you begin to write a review of literature, be sure you understand what constitutes plagiarism and how to avoid it. Document all information carefully, and follow a technical style sheet for specifications in textual citations and bibliographic entries.

Literature reviews bring together background information that links original research to related studies (see Appendix 2). They can serve to justify your work, support your results, establish your methods, or simply add to your knowledge on a subject. You will understand your own work better when you not only have read and studied what others have done but also have brought their ideas and yours together in writing. Literature reviews support proposals, theses, journal articles, oral paper presentations, and other reports. They may also be written to stand alone as review articles. Complete books, monographs, or journal articles can be made up of reviews of the literature. The suggestions that Day (1994) provides on writing review papers can be helpful in any literature review.

Writing the review is no easy task. Just as with the literature search, your efforts should be based on the objectives of your own research and the goals for your writing. Once you have searched the literature thoroughly, you read and become familiar with what you have found. Probably at this point you will begin organizing your notes. Think in terms of the organization that your review will take, and devise sections and headings based on your objectives. Your notes might, for instance, fall into categories such as "General Information, Methods in Other Studies, Support for Objective 1, Support for Objective 2, Similar Hypotheses, Results to Compare to Mine, History of Subject, Pros and Cons of Controversy, Also May Fit Somewhere." Note cards or loose leaves that can be organized and reorganized are handy for these notes, but if you prefer a bound notebook or the computer screen, you can develop a system for keeping up with the arrangement of your information.

After you have become thoroughly familiar with the literature on your subject, the job of writing a review of it may look formidable. Complex jobs are usually done best if you divide them into parts and attack each part separately. I have two suggestions.

Method A: Develop an Outline

Your subject matter will govern the kind of overall organization. If you are showing how your subject has developed over the years, you will choose a chronological arrangement. Comparison and contrast can be used in reviews of controversial theories on a subject. If your research brings together several views on a subject, you may enumerate and describe the various opinions; or you can organize on the bases of the topics involved in your own work. Whatever your organizational plan, have it in mind before you write. Better still, have it in mind and write it down too.

Look again at the headings you have given to your notes. You may now want to reorganize material, combine two parts, or separate one into two. In this reorganization, write down headings for the sections of your literature review. What you are doing, of course, is outlining.

Now, you can organize the material for each section to make independent outlines of small portions of the review. You may want to make outlines of all the sections before you begin to write in order to see how they separate or connect. Parts can be combined or separated and reorganized yet again. All this organizing and reorganizing may seem time-consuming, but it may well save you some time and help to produce a better paper.

When you are satisfied with the organization of a single section, write that one, then another, and another until you have a draft of each. Then go back and put them all together being sure you have made comfortable transitions for the reader.

Method B: Create a Skeleton

If outlines evoke from you screams of protest, do some freewriting first. Take a few of the documents you collected and write about what each of them says. For a fair-sized review as few as five or six articles will do. From this writing, you can discern the main points that are being made in the literature. Keep in mind your own objectives and discard anything that is not going to serve you well in your final draft.

Now, with the main points as well as your own objectives in mind, set up the sections of your review. Don't scream; you have to outline or at least organize sooner or later. The freewriting is for sorting things out so that they will start falling into place. Now the information you have freewritten from the few articles needs to be rewritten and arranged under the sections. This revision will serve as a skeleton of your first draft of the full review. You can take other documents that you wish to review and add or insert information from them into what you have written about the first five or six.

This method works well with the word processor in that you are revising and editing as you add and rearrange ideas from your literature. Word processing also helps you to sort out a stack of literature and get information from each piece into the right places in your text. You may wish to cite the same paper in more than one section of your review. With an initial skeleton of your work on the screen, you can add to one section and then to another as you glean information from a given article.

Methods A and B work quite well together. Whatever organizational technique your use, the draft will need to be revised to make the information flow to the reader in a logical fashion.

A common problem in writing the literature review is that it can turn into a boring list of ideas in paragraph form. Make yours move from one point to another with discussion and transition. Use subheads (not too many), transitional phrases, and unifying ideas to make information flow smoothly. Spice your writing with variety. Recall all the times you have almost fallen asleep trying to read boring material. Keep yours alive. Vary the way sentences and paragraphs begin. And be sure to use an assortment of active verbs. Literature reviews can quickly wear down a handy word. It is boring to read: "Author A found that. . . .Author B found. . . .What Author C found was. . . .Author D also found . . . ." Other verbs can substitute for *found.* Try some of the following as well as others: "Author A found, demonstrated, presented evidence for, suggested, observed, reported, examined . . . and concluded, noted, was convinced that. . . ." And vary the sentences by beginning with something besides the author's name: "In a study on . . . , Early in the 1980s, Author A . . . , According to Author A,. . . ." Our language is filled with active verbs and variations in sentence structure. Use them.

References

Day, R. A. (1994). "How to Write and Publish a Scientific Paper," 4th ed. Oryx Press, Phoenix, AZ.

Smith, R.C., Reid, W. M., and Luchsinger, A. E. (1980). "Smith's Guide to the Literature of the Life Sciences," 9th ed. Burgess, Minneapolis, MN.

5

THE PROPOSAL

"A new idea is delicate. It can be killed by a sneer or a yawn; it can be stabbed to death by a quip and worried to death by a frown on the right man's brow."

CHARLES D. BROWER

Your ability to write a research proposal can be vital to obtaining a degree, getting a job, and advancing to higher positions. As a scientist, you will encounter two important kinds of proposals: the graduate research proposal and the grant proposal. Distinctions in the two exist because of different audiences and purposes, but both require similar planning, composition, and execution. The graduate proposal is to provide a plan for you and your committee to follow relative to your research objectives. Your basic purpose in writing the grant proposal is to obtain funds to pursue certain research objectives. Developing skills in writing proposals is especially important for the young scientist who has not yet established a professional reputation. Writing the graduate research proposal can contribute to these skills.

A written proposal serves at least three purposes. First, it is your communication with a grantor, an advisor, or a committee. You are asking for approval and support to pursue a project. Second, when the answer is yes, then the proposal serves as an agreement between you and those granting the approval. The third purpose for a proposal is especially important to you. The contents become your plan of action and will serve as an outline of work to be done throughout the project.

Whether or not a written proposal, or prospectus, is required for a graduate research program, one should be written. For a graduate student, important reasons for organizing and writing out your proposal are that the writing helps you to plan your work in advance, to review what has already been done in the area, to foresee the pitfalls that lie ahead of you, and to remain focused on a reasonably direct track between your proposed objectives and your goals. A very practical side effect in writing a graduate research proposal is that some of it will serve as drafts for your thesis and for journal articles. With modifications, the introduction, review of literature, and methods will become parts of your thesis. Without a clear plan or written proposal, the time needed to acquire a degree can be increased by a semester, a year, or even more.

However, a graduate student proposal is not just a plan for you to follow; it is also a commitment to a departmental research program, to the graduate school that admitted you, and to your advisor. Your proposal is your commitment to make wise use of the time and resources available to you. Often in the scientific disciplines, the advisor, as well as the department and graduate school, has made an agreement with you to direct and support your research. If you fail to live up to your proposed commitment, you cost others time, effort, money, and progress in a research program. Your proposal gives those with interests in your accomplishments some assurance of success by presenting a goal to be accomplished, rationale and justification for pursuing that goal, and feasible methods to accomplish the goal.

FORM AND CONTENT

Whether your proposal is a graduate research proposal or a grant proposal, form and content will depend on your own intent and on the audience to which the plan is directed. Private and corporate foundations and government agencies that support research often supply a detailed format specifying the order and length of the sections in a proposal, information that must be included, and even specifications for the type, type spacing, and other elements of style. It is essential that you follow such guidelines. Many grants are competitive, and a proposal can be rejected simply because the author does not follow instructions. When a grantor receives 200 proposals and can provide grant funds for 10 of those, every detail is vital, even the appearance of the document.

Your graduate school, department, advisor, or granting agency may suggest a format or outline, but often you will have to devise your own organization. You can write a successful proposal by considering the characteristics common to all and setting up an organization that will best convey those characteristics to your particular audience.

Remember that communication is essentially a question–answer process. Producing a successful proposal requires knowing what questions your audience will ask and answering them effectively. In other words, first consider your audience and the following criteria that will be used to judge your proposal:

1. Scientific merit or benefit to the grantor
2. Importance to the discipline or the immediate problem
3. Feasibility
4. Rationale and methodology
5. Ability of the investigators
6. Budget and time required
7. Appearance and adherence to guidelines

Your proposal should answer several questions about a specific subject. Is it worthwhile? What are the chances of success? Are the investigators qualified to do the work? What benefits will be derived? Is the funding request realistic? The answers to such questions constitute the rationale or justification that serves as the heart of both the proposal and the subsequent research. Once you have identified the questions you are expected to answer, you are ready to decide what sections to set up to accommodate those answers. Almost any proposal will include at least the first six of the following conventional parts:

1. **Title page and executive summary or abstract**
2. **Purpose or hypothesis and specific objectives**
3. **Discussion of significance or need (justification)**
4. **Review of work done or being done (literature)**
5. **Materials and methods**
6. **Discussion of possible outcomes (conclusions)**
7. **Time frame, budget, and biography of investigator(s)**

With these parts in mind, you can begin to set up an outline for yourself or follow the format imposed on your writing. Again, the best organization for one proposal may not be best for another. You may begin with a discussion of the need for the research rather than your purpose or hypothesis; you can even discuss possible outcomes in the Introduction. You could start with a list of objectives and then build a case for pursuing them. The list above is **not** an organizational outline but a mere listing of what you should include.

A proposal for a graduate student's research may require a far more comprehensive literature review than the grant proposals would permit. However, you probably won't need the budget, time frame, and biography. You and your advisor or committee will have considered those questions relative to your program. But, you should become familiar with the format that requires these inclusions because they can be crucial for a proposal written to acquire grant funds in your future.

If you are not following a prescribed format, you should explore differing opinions of others before you begin. Stock (1985) and Smith (1984) have good summaries on preparing proposals and obtaining grant support. Basic strategies for writing proposals are discussed also by DeBakey (1978), Ebel *et al.* (1987), Hellweg (1979), and Meador (1985). Hellweg has good checklists in the appendices of her book. Regardless of what format you use, you should include the following parts in any effective sequence.

Title

The title should identify the specific subject in as few words as possible. It should attract attention to the hypothesis and clearly reflect the objectives of the proposal. Use only key words and avoid generalities or abstractions such as "A proposed study of the. . . ." We already know it's a proposal.

The title page is the first impression you make with your proposal. Be sure it is neat. In addition to a carefully worded title, this page will name the principal investigators (authors) and give their addresses, the date of submission, and the committee or agency to which the document is submitted. Many formats for grant proposals also require that the title page include information on the amount of funding requested and the time frame in which the work will be done. They sometimes require official signatures from not only you but also a company or institutional official who coordinates funding activities. Some groups, such as governmental agencies, provide a rather complex form to fill out as a cover page for the proposal. The cover form in Fig. 5-1 for the National Science Foundation is followed by a certification page that requires signatures of all principal and coprincipal investigators and legal statements or questions to answer regarding such things as contracts, cooperative agreements, debt, and disbarment, all of which you must also sign. If you have no guidelines, especially for the graduate research proposal, create your own neat title page (Fig. 5-2).

Executive Summary or Abstract

For the grant proposal, the executive summary or abstract is the most important impression you make beyond the title page. Make it concise but compelling. Its content is the basis for a decision to consider or reject the entire proposal. The overall neatness of the proposal, the extent to which it follows a

COVER SHEET FOR PROPOSAL TO THE NATIONAL SCIENCE FOUNDATION

FOR CONSIDERATION BY NSF ORGANIZATION UNIT(s) (Indicate the most specific unit known, i.e. program, division, etc.)	FOR NSF USE ONLY NSF PROPOSAL NUMBER
PROGRAM ANNOUNCEMENT/SOLICITATION NO./CLOSING DATE	

DATE RECEIVED	NUMBER OF COPIES	DIVISION ASSIGNED	FUND CODE	FILE LOCATION

EMPLOYER IDENTIFICATION NUMBER(EIN)OR TAXPAYER IDENTIFICATION NUMBER (TIN)	SHOW PREVIOUS AWARD NO. IF THIS IS: ☐ A RENEWAL OR ☐ AN ACCOMPLISHMENT-BASED RENEWAL	IS THIS PROPOSAL BEING SUBMITTED TO ANOTHER FEDERAL AGENCY? YES ___ NO ___ IF YES, LIST ACRONYM(S)

NAME OF ORGANIZATION TO WHICH AWARD SHOULD BE MADE:	ADDRESS OF ORGANIZATION, INCLUDING ZIP CODE:
AWARDEE ORGANIZATION CODE (IF KNOWN):	
NAME OF PERFORMING ORGANIZATION, IF DIFFERENT FROM ABOVE	ADDRESS OF PERFORMING ORGANIZATION, IF DIFFERENT, INCLUDING ZIP CODE:
PERFORMING ORGANIZATION CODE (IF KNOWN):	

IS AWARDEE ORGANIZATION (Check All That Apply): (See GPG For Definitions) ☐ FOR PROFIT ORGANIZATION ☐ SMALL BUSINESS ☐ MINORITY BUSINESS ☐ WOMAN-OWNED BUSINESS

TITLE OF PROPOSED PROJECT:

REQUESTED AMOUNT $	PROPOSED DURATION (1-60 MONTHS) months	REQUESTED STARTING DATE:

CHECK APPROPRIATE BOX(ES) IF THIS PROPOSAL INCLUDES ANY OF THE ITEMS LISTED BELOW:

☐ VERTEBRATE ANIMALS ☐ NATIONAL ENVIRONMENTAL POLICY ACT ☐ FACILITATION FOR SCIENTISTS/ENGINEERS WITH DISABILITIES
☐ HUMAN SUBJECTS ☐ PROPRIETARY AND PRIVILEGED INFORMATION ☐ RESEARCH OPPORTUNITY AWARD
☐ HISTORICAL PLACES ☐ DISCLOSURE OF LOBBYING ACTIVITIES ☐ INTERNATIONAL COOPERATIVE ACTIVITY: ______ Country/Countries
☐ BEGINNING INVESTIGATOR (See GPG SECTION I)
☐ GROUP PROPOSAL
☐ SMALL GRANT FOR EXPLORATORY RESEARCH (SGER)(SEE GPG SECTION II,C,12)

PI/PD DEPARTMENT	PI/PD POSTAL ADDRESS
PI/PD FAX NUMBER	

NAMES (TYPED)	Social Security No.*	High Degree, Yr	Telephone Number	Electronic Mail Address
PI/PD Name				
Co-PI/PD				
Co-PI/PD				
Co-PI/PD				
Co-PI/PD				

NOTE: THE FULLY SIGNED CERTIFICATION PAGE MUST BE SUBMITTED IMMEDIATELY FOLLOWING THIS COVER SHEET.

*SUBMISSION OF SOCIAL SECURITY NUMBERS IS VOLUNTARY AND WILL NOT AFFECT THE ORGANIZATION'S ELIGIBILITY FOR AN AWARD. HOWEVER, THEY ARE AN INTEGRAL PART OF THE NSF INFORMATION SYSTEM AND ASSIST IN PROCESSING THE PROPOSAL. SSN SOLICITED UNDER NSF ACT OF 1950, AS AMENDED.

NSF FORM 1207 (1/94) Page 1 of 2

FIGURE 5-1

Sample cover sheet for proposal to a federal agency.

Influence of Benzoic and Cinnamic Acids

on Growth and Survival of Heliothis zea Larvae

a proposal

submitted in partial fulfillment of the degree of

Master of Science

by

Gerald L. Bjornberg

to

Graduate Committee

Department of Plant Sciences

College of Arts and Sciences

University of Tokenburg 16 January 1996

FIGURE 5-2
Sample cover sheet for graduate research proposal.

prescribed format, and the content of this summary will determine whether yours makes the first cut in a group of competitive proposals. For those 200 competitive proposals submitted, the executive summary may be the only section read before the 200 are reduced to a lesser number to consider for funding.

The granting agency or your graduate committee may request an abstract rather than an executive summary. Used with the proposal, these two forms may be synonymous, but with either term what may be expected is simply a description of the proposed hypothesis, objectives, and expected outcomes. Try to acquire a copy of a successful proposal that has been submitted to your granting agency, or call personnel authorized to handle the proposals for the agency to ask for information. If you cannot obtain specifications, fulfill the request for an abstract or executive summary by following the characteristics for the informative abstract to whatever extent you can. The difference in this summary for a proposal and an informative abstract written for a report on completed research is that results cannot be included. This omission allows more space for justifications and methods. Begin with a sentence or two of justification followed by the objectives, a concise statement of the methods, and then the conclusions to reiterate justifications or benefits. Be sure your summary or abstract is well worded and establishes the credibility of the proposal and the investigators. The graduate proposal and short grant proposals sometimes omit this synopsis.

Introduction

The organization of the introduction will vary depending on the audience and the development of the full proposal. Whatever form it takes, it should immediately show the reader the subject to be investigated and give a rationale for pursuing the research. By way of identifying the relative scientific merit of the research and justifying its pursuit, the introduction may include some literature review and statement of benefits. It should include a purpose and suggest the scope of the proposed research. Above all, it should define the hypothesis and list the objectives.

Don't let the word **hypothesis** disturb you. As used here, the word simply refers to the proposition, the purpose of the study, the assumption you expect to prove, the question to be answered with the research, or the problem to be solved. It points toward what ideas can be credited or discredited when the objectives are satisfied. The **objectives** are specific goals. They should encompass the aims of the research, yet be brief, precise, and limited in number and scope. Trying to include too many primary and secondary objectives can obscure the focus of your proposal for you and for those who review it.

For clear justification and content in the proposal, be sure to distinguish between the hypothesis and the objectives in your own mind. Let me illustrate by following van Kammen (1987) in using Christopher Columbus as an example. If Columbus were writing a proposal for Isabella and Ferdinand, his title might be "An Alternate Trade Route to the Orient." His hypothesis would be that the world is round and that by sailing west he could reach the East, establishing a new trade route for Spain and the rest of Europe. His objectives might be (1) to sail west and chart a route to compare to other routes and (2) to bring home three shiploads of spices. Notice the differences in the hypothesis and the objectives. The hypothesis is the general supposition and contains a preconception not yet proven. The objectives are the specific goals that will be achievable, if the hypothesis is true, by acquiring empirical data (distance measurements) and tangible amounts of tea, cinnamon, and frankincense. These objectives have to be realistic and appeal to the grantor. Queen Isabella probably questioned how realistic they were, but both objectives would certainly be appealing to her. The same principles are true in developing a research proposal that appeals to an advisor and graduate committee or a funding agency.

Incidentally, while we are using Columbus as an example, let me call your attention to the observation that van Kammen (1987) makes about the "Columbus paradox." Christopher did not accomplish his objectives; his research hit a roadblock. He did not prove his hypothesis (although he thought he did), but his work made a definite impact on the Americas and the world. Further research on the same hypothesis did reveal a positive conclusion. You have not failed at all if your research leads you to the West Indies rather than to China. Your discovery may be more important than if you had been correct in your initial conception.

Justification

Justification is the key word around which a proposal is built. It is the basic criterion by which the final proposal is judged. Justification permeates the entire proposal from the title to the conclusions with an appeal for approval and with evidence that the proposition should be pursued. Whether or not a specific section is given this headnote, be sure that all sections adhere to the rationale that justifies the time, effort, money, and other support necessary to accomplish your objectives.

Justification outlines what can and should be done to accomplish beneficial outcomes. It shows how the methods can accommodate the objectives and how satisfying each objective will help to achieve the final goal. Justification describes the importance of your research to science and to the application of science. You can also justify your project in terms of its timeliness and economic significance as well as your own ability and access to resources to accomplish meaningful objectives. Whether it is a section unto itself or an integral part of all sections in the proposal format, the justification will be based on the following:

1. **Reason and logic**
2. **Scientific principles**
3. **Previous research (literature)**
4. **Feasibility**
5. **Use of or benefit from the results**

In some instances, your research may be partially justified on the basis of preliminary research you have done on a subject. If so, you may wish to present data you have collected, results that have come from your research, and a discussion of how and why these results indicate that further study has merit. These research results still need to be supported by scientific principles and the literature.

Literature Review

The literature review should consist of a summary of publications pertinent to your research. It can review the history of your subject or the present state of the art; it ought to consider any controversies surrounding a question or gaps in available information that you intend to fill with your research; it may introduce methods that will make your work possible. Whatever kind of support it gives your proposal, a good literature review can establish your credibility and your chances for having a proposal accepted by illustrating that you know what has been done, what is being done, and what needs to be done in an area. You should cite work being done and recent publications from other institutions as well as from your own to establish the relationship between your proposed research and that of others. Be sure that all discussion is relevant to the specific objectives and the general hypothesis. For most proposals, keep all sections, including the literature review, brief to hold the reader's attention to your objectives. Some advisors expect a lengthy review of the literature in a graduate proposal.

Methods

In the review of literature or in the methods section, you will increase your credibility if you point out methods that other researchers have used with or without success. The methods section, often called "Plan of Operation," "Materials and Methods," or "Experimental Procedures," is the very foundation of the scientific merit and feasibility of the work. To convince your audience that your plan is feasible and to serve you in pursuing the research, the methods section should outline the working plans in as much detail as possible. Include information on materials, sampling, analysis, data, and even people you will need to work with; steps you will take in conducting the research; data you will collect; and how you will analyze and use the data collected. Describe any potential problems you may encounter and tell how you will address these. Procedures should generally follow the same order as the objectives and show how each objective will be attained. All materials and time needed should be justified in this section and reflected in the budget of a grant proposal. To the granting agency, this section should justify the budget.

Discussion and Conclusions

Although you may have no results and little discussion beyond that described in previous sections, it is important to reemphasize objectives, summarize points in the justification, and draw the reader back to the research question, the hypothesis, and the objectives. Your conclusions will enumerate points of justification and benefits to be derived. This section can extend into proposed applications or future research beyond your own, but don't overdo this idea.

References

The reference section is essential to your proposal. References with full titles indicate the extent to which you have explored your subject and are helpful to reviewers in their considerations. The citations and references must be accurate and follow a consistent style throughout. Any error can destroy your credibility and diminish the chance for acceptance of your proposal.

Budget and Time Frame

For a grant proposal, the most important issues can be time and money. A grantor may be fully convinced that your scientific proposition is feasible, but an unrealistic budget can negatively influence opinions of reviewers. Just as some people ask for too much time and money, some do not ask for enough. Be realistic in your needs.

Plan the budget carefully with estimates as nearly accurate as possible. You don't know exactly what your supplies will cost, but if your research has been clearly outlined, you can provide a reasonable estimate. Include salaries, equipment and supplies, publication costs, travel to collect samples or to take results to professional meetings, and even phone bills. Don't forget indirect costs or overhead if your company or institution charges such a fee. Many times overhead is a large percentage of the total costs; 20%, 40%, and even higher percentages for overhead are not unusual. On the other hand, some agencies may not allow such indirect costs to be budgeted. Be sure your budget can reflect an agreement between your employer and the funding agency.

Some grants are made only if the employer matches the value of grant funds in money or other resources. In this situation, you will need to include in your proposal a statement on these contributions. Describe the facilities and equipment that your employer will contribute to carry out the project. Also note the percentage of time the investigators and other staff or hourly workers will spend with the project and consider percentages of these salaries as part of the employer's contribution.

Time may be as important to your grantor as money. Allow yourself enough time, but don't be wasteful. In asking for time, justification in the body of your proposal is critical. Some experimentation takes longer than other. If you are doing laboratory analysis that takes just 5 weeks for each data set and you are supplying 10 data sets, you might reasonably complete your proposed research and all the reports involved in 1.5 to 2 years. However, if you are doing field work that must be repeated over the environmental conditions of 2 or 3 years, you can justify additional time. With both time and money, the key word is realistic.

As a graduate student, you may not need to worry about your budget and time frame. Your advisor does enough worrying for both of you. You simply accept the support available, and your proposal concentrates on what you will do rather than how it will be financed. The time frame may be dictated to you. At some schools, you must finish your program within a certain time, or your research support will no longer be available to you. Consequently, you work within the time and money limits provided.

Biographical Information

For your graduate proposal, you will not need to provide biographical information because your advisor has already looked into your credentials with your application to graduate school. But for the grant proposal, what is often known as **curriculum vitae** is usually essential. Be sure to make clear in this abbreviated resume that you are capable of doing the proposed research. Emphasize points in your training that have to do with the expertise needed and avoid an extensive presentation of points unimportant to the research.

OTHER CONSIDERATIONS

Proposals are likely to become a vital component in your career as a scientist. Be conscious of the requirements for a good proposal as you write your graduate prospectus. The practice you get will serve you well when you begin to write grant proposals to support your own research. In graduate school, as a graduate asssistant or a research assistant, you may assist in writing grant proposals other than that for your own research. For example, toward the end of your graduate program, you might be asked to assist with a proposal extending research beyond the limits of your own program or pursuing another avenue of research in the same discipline. Producing any proposal can be a valuable experience for you, and it will teach you much about the proposals you will produce in the future.

Despite all the work you put into writing proposals, many of them will be rejected. Reduce your frustrations by recognizing the beneficial side effects to writing unfunded grant proposals, or even the graduate research proposal that your committee does not approve. Something positive often comes despite the rejection. Think of it this way. You have a good idea for scientific research. In writing the proposal, you pursue the subject in depth. You learn much from the literature; you learn who else is working in the area of research and often get acquainted with people you'll work with in the future; and you practice your skills in writing proposals. In the face of a $75,000 rejection, these benefits may seem trivial, but in the whole picture they are important to your career. Proposal writing is a way to keep up with what is going on; it prompts you to read the literature and to consult and cooperate with your colleagues.

As with any other important document you compose, your proposal should go through a process of reviews and revisions. Often you must meet deadlines for submission of proposals. Write drafts in plenty of time for reviewers to make suggestions and for you to revise the proposal.

Whether a proposal is invited or competitive, be neat and direct, follow guidelines, and be realistic. Your ability, the time needed, the resources available, and the money granted must all add up to probable success from a proposed project. If these things are not treated realistically, all the good scientific ideas in the world will not yield as many positive side effects nor as much funding for your research.

Finally, as important as obtaining the approval or funding for your research is the execution of the project. As you get into the work, you may find that you have to revise the proposed plan. For instance, you will find a method that provides better results, you will discover evidence that must alter your original hypothesis, or specific equipment will become inoperable just when you need it. Your original proposal will not guide you through all the possible problems and discoveries that occur. You must be creative enough to find a new path and execute the research regardless of obstacles.

To keep their records updated and because the unpredictable is to be expected, many grantors and graduate advisors require periodic progress reports. These reports protect your credibility. Be sure you get the reports in on time and clearly explain the state of the project, what has been accomplished, and the proposed next steps. If you've had problems, don't be too pessimistic in these reports; be positive, but as with all the details in your proposal, be realistic. Grantors and advisors are generally reasonable people. They need to know that you are doing the best job possible in carrying out your proposed project.

Progress reports can also help you with that final step in any research project—the publication or thesis. No research project is complete until this final report is done. This report is evidence of your accountability or evidence that the time, money, and resources have been used well. Among the characteristics that are scrutinized in a beginning scientist is the potential for good grantsmanship or "the art of getting financial support (a grant) for your research" (Stock, 1985). Your ability to communicate through the proposal and in a final thesis, report, or publication is an important component in this art and absolutely vital to your career as a scientist. An example of a graduate student proposal is in Appendix 3.

References

DeBakey, L. (1978). The persuasive proposal. *In* "Directions in Technical Writing and Communication" (J. R. Gould, Ed.), pp. 25–45. Baywood, Farmingdale, NY.

Ebel, H. F., Bliefert, C., and Russey, W. E. (1987). "The Art of Scientific Writing." VCH, New York.

Hellweg, S. (1979). "Techniques of Grantsmanship." Institute of Public and Urban Affairs, San Diego State Univ., San Diego.

Meador, R. (1985). "Guidelines for Preparing Proposals." Lewis, Chelsea, MI.

Smith, R. V. (1984). "Graduate Research: A Guide for Students in the Sciences." ISI Press, Philadelphia.

Stock, M. (1985). "A Practical Guide to Graduate Research." McGraw–Hill, New York.

van Kammen, D. P. (1987). Columbus, grantsmanship, and clinical research. *Biol. Psych.* **22,** 1301–1303.

6

THE GRADUATE THESIS

> "I was taught that the way of progress is neither swift nor easy."
>
> MARIE CURIE

THE THESIS AND YOUR GRADUATE PROGRAM

Except for works of rare genius, a thesis or dissertation (I use *thesis* to mean either) cannot be produced in a week or even a month. No designated time can be set for work on the thesis—some take 6 months, others a year. Preliminary work can add another 2 years. To make the most conservative use of time, the thesis should be written as the graduate program progresses (see Appendix 4). This time must be coordinated with other activities such as research, course work, and professional meetings. If you find that it's hard to juggle all these activities, read Molly Stock's (1985) book, *A Practical Guide to Graduate Research.*

Characteristically, your thesis should be built on

1. *A complete library search* on everything that has been done on the specific subject and closely related subjects.
2. *Original research or a professional project*—Field and laboratory experiments based on a research proposal or project as approved by your major professor and graduate committee.
3. *Your syntheses*—Putting together and deriving meaning from data, ideas from others, and your own conclusions.

To neglect any one of these foundations for your thesis can limit the quality of your work. Success depends on your knowledge of what others have concluded from their studies, meticulous work with your own research, and your efforts to give a clear, accurate perspective to the entire study. When these points have been given appropriate attention, the other criterion for success is communication—the writing itself.

The master's and doctoral theses differ a great deal, and expectations from one graduate school to another are inconsistent even in the same discipline. Specific requirements are set forth by the department, the college, and the graduate school from which you get your degree. As you begin your graduate project, talk with your advisor and peruse several theses that have been produced in recent years by reputable degree candidates in your department. In other words, get a clear feel for what your thesis should be long before you write it. Typically, the thesis will include the following:

1. ***Introduction*—General justification for the study, the hypothesis or purpose behind the study, and a specific statement of objectives.**
2. ***Literature Review*—A detailed report from your library search about what has already been done on your subject (sometimes combined with the introduction).**
3. ***Materials and Methods*—An account of the specific procedures or techniques used in the study including statistical designs and analyses.**
4. ***Results*—A presentation of the data acquired from your research.**
5. ***Discussion*—Significance of your own data as well as the relationship between your work and the findings of others (the Results and Discussion sections may be combined).**
6. ***Conclusions*—A summary of your findings and their significance as well as suggestions for further research or applications for the findings.**
7. ***Bibliography*—References or literature cited.**
8. ***Appendices*—Related materials that support a point and provide additional information but are not essential for understanding the thesis itself.**
9. ***Abstract*—Doctoral dissertations usually require an abstract, and one may be needed for the master's thesis.**

Writing the thesis will be easier when you visualize a clear picture of the content and organization involved. As with your proposal, the content of your thesis will begin with presentation of a question to be answered or a problem to be solved. You will establish objectives for your study by reviewing the literature and suggesting a hypothesis for answering the question. If your proposal is carefully written and followed, it can become a foundation for the introduction, literature review, and methods sections of the thesis. Once you have tested your hypothesis with your methods, you can report and interpret results relative to the original question. With this vision of your thesis in mind, consider use of the following resources.

Graduate College Requirements

Most graduate schools furnish information on requirements in a catalog and in a guide for preparing theses and dissertations. Keep these instructions handy. There are deadlines to meet and fees to pay. Thesis requirements include time available, committee composition, and technical details such as margins, typeface, spacing, and the kind of paper required. Knowing these things ahead of time will help you avoid problems later.

Style Sheets

For points of style beyond those specified by your department or graduate school, the discipline in which you are working probably has a style to which it generally adheres. For example, *Scientific Style and Format* (CBE, 1994) is a convenient reference on points of style such as abbreviations, punctuation, and bibliographical style. If you don't know which style manual to consult, ask your advisor. Sometimes your advisor or committee will recommend that you choose a professional journal in your discipline and follow its style. Styles for publications differ with journal editors or publishers, but most provide "Instructions for Contributors" or other style sheets to follow. Become familiar with the style of your professional societies and have a style sheet handy.

The Library

The sooner you get acquainted with a library, the more time you will save yourself. The simplicity or complexity of your literature search will depend on your knowing what you want to find and how to find it quickly.

Your Advisors

Your major professor is probably your most valuable resource. Take advantage of his or her expertise. Report to that advisor regularly, but don't make a nuisance of yourself with questions that can be readily answered with a graduate school catalog or your style manual. Because departments and graduate divisions differ in their requirements for theses, no handbook can provide the final word for what your thesis should be. Your major professor can. In addition, your other committee members can be valuable consultants as you proceed with your study. Each is on your committee for a particular reason. Get acquainted with them early, and visit with them periodically.

Other Professionals

Unless you have clear expertise in all areas of your project, you may need to consult specialists in addition to your major professor and other committee members. For example, your thesis will most likely contain quantitative data, and a statistician should be consulted before you plan your experiment and collect the data. Be able to furnish him or her with an outline or thesis proposal that includes your research hypothesis and objectives. Formulating the design for your experimentation and stabilizing your plans early in your research can pay big dividends in both time and research quality later.

AVOIDING PROBLEMS

Your thesis should be the written record of your graduate research project and contribute substantially to your professional reputation. Building and maintaining your reputation with your peers and faculty will depend on not only how good the final product is but also how you handle problems along the way. The proposal, your research, cooperation with others, and the thesis will illustrate your scientific ability and professionalism. How well you integrate these individual activities will determine, in large part, how successfully and quickly you complete your degree. Weeks, months, even years of delay can result from poor planning and execution of the graduate research project and thesis writing.

Work closely with your major professor but assume full responsibility for your program. Don't wait for the professor to tell you to write a proposal, search the literature, and write a literature review. If these chores are not required by your department, you'll be ahead if you do them anyway. If the proposal is not required, write one and take at least an outline of your plans to your professor and ask for his or her advice. Then suggest that you submit the entire proposal to your committee and meet with them for their opinions. Your major professor will likely be pleased that you are assuming responsibility.

However, don't step across the line of diplomacy. The department has policies, and you are probably using resources that belong to your department and your major professor. Consult with your advisor before taking drastic steps. As you become acquainted with departmental procedures and the personalities you work with, you will be able to determine how much independence you have. The following suggestions can help you avoid pitfalls common to graduate students.

Get Started Early

The responsibility for getting the thesis finished is yours alone. From the day you begin a graduate program, planning for your thesis begins. Decide very early what area you want to work with so that your advisor, your course work, your exploration in the library, and your research in the laboratory or field can be chosen with the specific objectives of your thesis in mind.

Maintain a Professional Relationship with Your Advisors

Recognize that advisors are humans with unique personalities. You are probably not the primary focus in their work, but they should be actively involved with your project. Don't be offended when they don't let your project take precedence over their many other activities. If there are times when your major professor is too preoccupied to help you, take charge of your own destiny, with finesse and diplomacy of course. Work cooperatively, but do some independent thinking too. That's part of being in graduate school. If you assume full responsibility for your program, you should finish your degree on schedule regardless of how much or how little input the advisor contributes.

Professors are abused in two ways: You ask too much or you ask too little. No professor has time for you. You waste their time with questions you could answer by consulting a dictionary, a college catalog, or a style manual. You waste their time with bits and pieces of your thesis that are too hard to read. You waste their time with too much casual talk or too many intrusions into their work schedule. However, a casual question or remark, a handy reference you find at the library, a personal revelation—these things can be very important to the advisor's knowing you and your subject. Be discreet in how much time you take.

Graduate students also abuse their professors by avoiding them. Paradoxically, they have plenty of time to work with you. Your work is important to their own. Consultations on courses they teach, subjects of common interest, your thesis—all are important for both of you, and the advisor does not feel that such things are a waste of time. Certainly the research for your thesis and the writing of it are points that bear repeated discussions. A student who writes a mediocre thesis and dumps an error-ridden copy onto an advisor's desk 3 weeks before expecting to graduate is certainly abusing the advisor and jeopardizing graduation.

Draw Up a Carefully Planned and Well-Written Proposal

An attempt to pursue graduate research and produce a thesis without a written proposal is like taking a trip through unfamiliar territory without a road map.

You may find your way and even be successful, but most likely you will waste time, have to double back over some roads, go down blind alleys, and even get lost. Working out hypotheses, objectives, justification, literature, and methods for the proposal will sharpen your perception of your subject. The written proposal provides an early draft or outline of the thesis, and it will serve both you and your graduate committee in communicating and keeping on track.

Maintain Accurate, Complete Data

All data you collect should be considered important. Gather them carefully, write them legibly, analyze them thoroughly, respect their revelations, and then store all of them, not just the part you use. Read Macrina (1995) on keeping scientific records.

Don't trust your memory. Write down field plans. Write down chemical analyses. Write down techniques you use, amounts you use. In field observations, note (write down) the weather. Record full references from the library with page numbers and full names (*et al.* can get you in trouble). Your good memory cannot always fill in the blanks.

In addition, for many of you a camera can help record data. Pictures you take can be important in showing results of your research and can often be used in the thesis and certainly in slide and poster presentations. Keep in mind that, at present, most journals publish only black and white photographs. You may need to record some information on black and white film. Images on most color film can be reproduced in black and white, but quality will be lacking unless you use super-fine-grained, slow-speed color film. Have access to a camera and use it often.

Write the Thesis as Your Work Progresses

It would be a formidable task to write the whole thesis at one time. But you can divide the work into logical portions. Before you begin your research or when you write the proposal, you should make a complete library search and write a rough draft of the **Literature Review.** At this point you can compile a first version of your **Literature Cited.** When you set up your proposal and plans for research, you can also write the **Methods.** As portions of your research are finished, you can draft the **Results and Discussion** section(s). All of these sections will require revision, but the easiest time to write each first draft is when details are fresh in your mind. Your **Introduction** and **Conclusions** can be the last sections you write. The introduction may be a modification of the introduction to the proposal. The purpose for the introduction is to direct the reader into the thesis; the conclusions need to focus the reader's attention on the most important findings. You can best accomplish these purposes after you see where you have been.

FIGURE 6-1
Don't bypass the advantages offered by the computer for processing words and data.

Use a Word Processor

If you have not developed the capability of composing at a keyboard, now is a good time (Fig. 6-1). Don't bypass the advantages that are offered by computer processing of words and data. Writing, designing tables and graphs, revising, editing, and printing can all be made easier with the word processor. If you are just becoming familiar with word processing or if you have computerphobia, read Zinsser (1983).

Make the Final Copy One That You Can Be Proud Of

Because it is a reflection of your reputation, you should not skimp on materials or quality for your final copy. You may have to pay for such services as typing, making figures, producing photographs, and providing 100% bond paper. These things can be costly; be alert to how to cut costs, but be sure the final product is a reflection you want. Others may contribute to it, but your name stands alone as the responsible author.

Finish before You Go

The last weeks of your graduate program can be hectic. In addition to getting the thesis in a final form and completing other requirements, you probably will also be looking for a job. Job offers often come before your thesis is ready to turn in. If at all possible, arrange to begin work only after the thesis is complete. The new job will require your full attention, and even if you have some time in the evenings, you will find it hard to focus on the thesis. The psychological and academic influences of being near your major professor, the library, and other graduate students are important to your maintaining a focus on the thesis. Former students who have tried taking the incomplete thesis to a job repeatedly implore me to advise other students not to do the same. Finish before you assume the next job.

To Publish Is to Build Your Reputation

The best time to publish is when your research and data are fresh in your mind. Sometimes a doctoral dissertation and occasionally a master's thesis can be published as a book or monograph. With or without direct publication, you may want to copyright your work. You can do that yourself or you may operate through a publisher. If your work is a doctoral dissertation, your graduate school will probably require the publication of an abstract with Dissertation Services (UMI Dissertation Services, 300 North Zeeb Road, Ann Arbor, MI 48106–1346). That agency also offers other publication services such as on-demand publishing, obtaining copyrights, and binding.

Whether or not you copyright or publish your entire thesis, you may want to consider publication of your research results as journal articles. From your thesis, you can choose the most significant data and arrange them to agree with the format of a given journal. It is relatively easy to put a paper together while the data are new. If you postpone this effort, you often never carry it out simply because the task becomes increasingly difficult as time passes. You will also have new goals to accomplish that do not allow time for going back to old data. Without extension of your materials into an active journal, much of your valuable data can grow stagnant in your thesis on a library shelf. Some advisors now suggest that the thesis itself be styled to the form of a journal article or articles. Seriously consider this possibility and work toward it from the beginning of your program.

Last-Minute Jobs

After your thesis is written to your own satisfaction, final chores will likely take several weeks. In preparing for your final committee meeting or defense of your thesis, be sure to get copies to committee members at least 2 or 3 weeks in advance to give them plenty of time to read and evaluate your written work. You may also want to visit with each of them individually for specific suggestions before the final meeting. New suggestions invariably come from individual committee members and their joint considerations. Allow time after their reviews to change entire sections and particular details.

After you have polished your thesis by incorporating the suggestions from committee members, typically you take to each of them your letter-perfect thesis along with three or more copies of a signature page for final approval. You may be lucky. You may get the signatures in 2 hours. However, 5 days later you may be still trying. One committee member is out of town collecting research specimens, another is at a meeting in New Orleans, and a third just happens to be out of the office every time you check. Still another committee member would like time to look over the revised thesis before signing it. That person is not being disagreeable nor picking on you. Advisors should look at what they are signing. Their reputations, as well as yours, are at stake. Although almost any student wishes for quick approval of the final thesis, serious students know that their own degrees and reputations are far more highly respected if permissive is not the key word in the reputation of the department from which they graduate. The professor should check your thesis a final time and even ask for further alterations in text, if needed, before signing the approval page.

The last minute has now evolved into weeks, and you still are not finished. Double check your graduate school instructions for special requirements such as paper quality and margins. Final copies must be collated with no page missing. Even the best equipment or printing service sometimes errs. When you deliver copies to your department, the library, or the graduate office, your thesis can be rejected if you have not been careful enough with details.

Along with the thesis you may need copies of a copyright release that you have to sign to allow library personnel to reproduce your work for research purposes. Because this signature must be yours, don't leave your thesis with a friend to hand in without signing the forms. Obtaining original signatures, even your own, can take days and cancel your graduation if you are pushing deadlines to their outer limits. With your thesis delivered to the designated offices, you are almost finished. You should offer a copy to your major professor and even to committee members who have special interest in your subject. Keep a copy for yourself. You will probably not want it in sight for some time, but some day you may even display it—at least on a bookshelf at your home or your office.

PLANNING THE THESIS

A first step in planning your thesis is to determine the form it will take. Your choice of form partly depends on the content of the thesis and whether or not the work can be published as a monograph or as journal articles. The choice also depends on what form is acceptable to your graduate school, department, and major professor.

In reference to doctoral dissertations, the Council of Graduate Schools (1991) found that the "traditional" thesis is "alive and well at all universities participating" in their study and this form "continues to represent the model in all fields." The Council recommends "flexibility with respect to form." However, the "consensus is that simply binding reprints or collections of publications together would not be acceptable as a dissertation. . . . Dissertations . . . require a fuller review of the relevant literature and a more complete discussion of results and conclusions than a journal would allow."

Gaffney (1994) has edited a policy statement by the Council of Graduate Schools on the master's degree, but that statement has little to say about the thesis itself. Requirements for the master's degree differ even more than those for doctoral programs. Some schools do not require a thesis, and others involve other capstone experiences such as a performance, extra hours of specialized course work, or other special projects. In the sciences the thesis is often preferred. What the Council does say about the M.S. theses indicates that we can adapt standards for the doctoral thesis to the master's level. The Council contends: "A master's student who does a thesis or project should be required to design the research project with the help of a faculty advisory committee, conduct the necessary background literature search, do the research, analyze the results, write the thesis, and communicate the results at an oral thesis defense." Unlike the requirement for a doctoral dissertation, the Council suggests that the master's work "will not necessarily be original research, but it will be a new application of ideas."

These standards set forth by the Council of Graduate Schools do not designate forms acceptable for either the M.S. or Ph.D. thesis. Departments in the life sciences commonly accept, or some even prefer, the incorporation of journal articles into both theses and dissertations, but the quality and standards established with the traditional thesis should not be sacrificed at either level. Substituting publishable journal articles for a traditional thesis or dissertation offers the distinct advantages of providing the graduate student with experience in writing for publication and of publishing concurrently with acquiring the degree. Although the Council of Graduate Schools (1991) contends that the journal article alone does not fully encompass the traditional requirements for a graduate thesis, literature reviews and complete reports on data collected can be incorporated into a thesis along with review-ready or published journal manuscripts. You need to understand what is expected and how to arrange the supplementary materials with the journal articles.

ACCEPTABLE FORMS FOR THESES

What, then, should be put before graduate committees and be acceptable for a master's thesis or doctoral dissertation? Because that question has no specific answer, your job is to decide what form you prefer and be sure your graduate school and advisors will approve your choice.

The same basic outlines for traditional theses or those incorporating journal articles will fit both the master's and the doctoral levels if a clear understanding exists regarding the differences and the educational requirements for each. Consult with your major professor before you decide which format to pursue. Consider the following possibilities.

The Traditional Thesis or Dissertation

Basic outline

Introduction ⟩ **May be combined**
Literature Review

Materials and Methods

Results ⟩ **May be combined**
Discussion

Conclusions

Bibliography

Appendices

Samples of well-done traditional theses are available in any university library. Until the 1970s this form for the thesis was expected from almost all degree candidates, and you and your major professor may still prefer it. Also, with exploratory research suitable for a master's degree but with data inadequate for publication, we can simply use this model, require a thorough literature review, a clear presentation of all procedures and results, and perhaps suggestions for additional research which may ultimately lead to publishable information.

We expect that the traditional thesis will follow the outline above, but reasonable creative alterations are not only acceptable but often commendable. Some schools may recommend length, but keep in mind that, for any thesis format you choose, the length of the thesis is essentially irrelevant. Some studies will require a 200-page report; others would be padded if they went far beyond 50 pages.

Theses or Dissertations Containing Journal Articles

The thesis styled with journal articles will be a compilation of manuscripts, which can be excerpted to be published, as well as sections such as the literature review and conclusions that are important to your entire study. If this thesis format is planned from the beginning of your study, much of the writing can be done before your research is complete. The literature review and its bibliography will need only to be updated at the conclusion of your program. Each journal article can be partially written as you recognize which objectives and methods will be included.

Basic outline

Preface (optional)

Introduction

Literature Review — **May combine** (Introduction and Literature Review)

Journal Article I—style of a specific journal

- **Abstract**
- **Introduction**
- **Materials and Methods**
- **Results and Discussion**
- **Literature Cited**
- **Tables and figures**

Journal articles (*ad infinitum*) journal style

Complementary Studies—for auxiliary, preliminary, or supplementary work not suitable for journal publication

Conclusions

Bibliography

Appendices

The Master's Thesis Including Journal Articles For the M.S. candidate who has produced findings for a journal article but has acquired supplemental material that should be recorded in a thesis, I suggest the following possibilities. (The thesis will not necessarily be written nor arranged in this order.)

1. Write a complete review of the literature when your proposal is designed and before your research is done. It will become a separate chapter in your thesis.
2. After your study is complete or as you complete the analysis for data sets, write the journal articles and note what you have omitted from those manuscripts.
3. Write a complementary section on auxiliary studies or supplemental data collected that are not included in the journal articles. This portion of the thesis may appear in the form of a manuscript you do not expect to publish, as a report or series of reports, or as appendices. The form depends on what remains to be presented.

4. Write comprehensive conclusions to your entire study. If the complementary section is included as appendices, the conclusions may precede that section. These conclusions will be commentary on the extent to which you have accomplished your objectives and satisfied the initial hypothesis. They may contain suggestions for further research and suggested applications for your findings. Theses will differ a great deal in how extensive the conclusion section should be, and with discussion in other sections, some theses may not benefit from additional conclusions.

5. Rather than a complete bibliography at the end of the thesis, place references or literature cited at the end of each chapter. At the end of the literature review, you will have a list of all citations contained therein. Literature with each journal article will be presented in the respective chapters in the style for the designated journal. This technique requires duplication of citations from the various chapters, but it allows for review-ready copy for the manuscript. Similarly, numbering of tables and figures can be by chapter rather than sequentially throughout the thesis.

6. Your thesis may need a preface to explain to the reader the arrangements you make for style, organization, and content within the thesis. If the explanation is not complex, it may be put into the introduction.

Doctoral Dissertation Incorporating Journal Articles Again, the Ph.D. candidate must do all the M.S. student has done and more because original research and creative ideas are expected. The doctoral dissertation is likely to be made up of more than one journal article and to contain additional supplementary materials. If the philosophic requirements for your degree are to be met, you must present a more extensive conclusion with commentary on scientific theory and philosophy that could not be published in journal articles. In addition to the same basic suggestions proposed for the master's thesis, I recommend the following for the doctoral dissertation.

1. Even after you have written two or more journal articles, you will have a great deal of supplementary material left over. You may need to consider presenting those data in more than one chapter as well as in appendices.

2. You will review the literature more extensively and include ideas from related published works that may not have direct reference to your own objectives but do support the scientific principles behind your objectives.

3. As with the review of literature, the conclusions section to your thesis should treat your subject more intensively and extensively than would the master's thesis. Discussion and conclusions will demonstrate an ability to think creatively and to integrate scientific ideas from the literature with your original research.

In other words, your thesis as a doctoral candidate is a *dissertation,* i.e., a lengthy treatise dealing with a subject from several perspectives that can direct your own scientific thinking and stimulate that of others in order to make a significant contribution to the scientific community. It is a document that helps to demonstrate your qualifications for the doctor of philosophy degree.

Caution: If you publish articles that you intend to include in your thesis before the thesis is complete, you may have to transfer copyright privileges to the publisher. Be sure to acquire an agreement with that publisher to give you the right to publish the same article in your thesis.

Theses Containing Journal Articles and a Proposal

A thesis or dissertation may also be produced with the inclusion of journal articles along with a proposal. The proposal would be an initial chapter in the thesis and might contain the comprehensive literature review. More likely the literature review would be a separate section coming before the Proposal section. These sections would be followed by the journal articles and whatever complementary chapters, conclusions, and appendices were necessary.

Basic Outline

Introduction—Might be in form of preface

Literature Review—A comprehensive review

Proposal

- **Introduction with justification and objectives**
- **Background studies—If these include a comprehensive review of the literature, the Literature Review chapter may be omitted.**
- **Materials and Methods**
- **Conclusions—Reiterating justification and benefits of the study**

Journal Article I—In the style of the journal

Journal articles (*ad infinitum*)

Complementary studies—As described previously

Conclusions

Appendices

Suggestions regarding the thesis or dissertation with journal articles are also applicable to this form. Under this technique a well-written and comprehensive literature review done in conjunction with the proposal would need only to be updated at the end of the study. The proposal itself would outline all studies that constitute the proposed graduate program. The conclusions section would bring together the proposal and the journal articles and accommodate any discrepancies between what was planned and what was actually accomplished as recorded in the journal articles and complementary chapters.

One concern with this form for the thesis is that the proposal may need to be modified as the research progresses. Almost any extensive graduate study must evolve from an original plan, but new discoveries may be made or new methods discovered that require you to deviate from the original proposal. Because the proposal is needed to establish the course of study, changes made along the way need not discredit the original plan nor create a need for rewriting it. A record of the entire graduate program can be seen in the thesis that begins with a proposal, is carried through reports on the research, and ends with a discussion of the successes and failures and their significance.

Background information on scientific theses is good in Stock (1985), Smith (1984), and O'Connor (1991), but most of the recent references that focus solely on theses have emanated from the humanities and social sciences with reference to the styles of the Modern Language Association or the *Chicago Manual of Style*. With this focus, they essentially ignore the unique characteristics of the scientific thesis. They do contain good general information, however. The one I'd recommend first is the guide by Mauch and Birch (1993).

Caution: Be sure that you have the approval of your advisor, committee, and graduate school before you decide on the form for your thesis. The ideas expressed here are not necessarily new, and certainly not revolutionary, but many advisors or graduate schools have rigid criteria that may not be met with these suggestions.

References

Council of Biology Editors (CBE) (1994). "Scientific Style and Format: The CBE Manual for Authors, Editors, and Publishers," 6th ed. Cambridge Univ. Press, Cambridge, UK.

Council of Graduate Schools (1991). "The Role and Nature of the Doctoral Dissertation: A Policy Statement." Council of Graduate Schools, Washington, DC.

Gaffney, N. A. (Ed.) (1994). "Master's Education: A Guide for Faculty and Administrators." Council of Graduate Schools, Washington, DC.

Macrina, F. L. (Ed.) (1995). "Scientific Integrity: An Introductory Text with Cases." ASM Press, Washington, DC.

Mauch, J. E., and Birch, J. W. (1993). "Guide to the Successful Thesis and Dissertation." Dekker, New York.

O'Connor, M. (1991). "Writing Successfully in Science." Harper Collins *Academic*, London.

Smith, R. V. (1984). "Graduate Research: A Guide for Students in the Sciences." ISI Press, Philadelphia.

Stock, M. (1985). "A Practical Guide to Graduate Research." McGraw–Hill, New York.

Zinsser, W. (1983). "Writing with a Word Processor." Harper & Row, New York.

7
PUBLISHING IN SCIENTIFIC JOURNALS

"In science the credit goes to the man who convinces the world, not to the man to whom the idea first occurs."

SIR WILLIAM OSLER

Almost all landmark publications, such as that of Watson and Crick on the structure of DNA, are the result of volumes of reports preliminary to the momentous discovery. Perhaps every scientist dreams of making a groundbreaking discovery in research and of publishing an article that will be considered the classic of the discipline. Not only must the discovery be outstanding, but also the article must be a well-written, clear disclosure. Not everyone can be that one-in-a-million scientist to provide the world with a momentous revelation, but you certainly can provide reports that serve as the stepping stones toward that classic publication. The most common vehicle for communicating with other scientists is the scientific journal. Contributions to the journal literature can go far in building your professional reputation.

When data are collected and analyzed and the results of your research are ready to publish, you have several decisions to make. When can you finish the writing? Will you have coauthors? Who will give you good reviews? Which journal will you submit your manuscript to? How soon will it be published? What is the possibility for acceptance? How do you deal with editors? Most of these questions do not have simple answers, but some general ideas may help you to follow the process of writing for journal publication.

PLANNING AND WRITING THE PAPER

Before you write the paper, determine who your coauthors will be and your position with them. Read about giving credit in *On Being a Scientist* (Committee on the Conduct of Science, 1989), and check what Bishop (1984) or style manuals such as the *CBE Style Manual* (CBE, 1983) have to say about multiple authorship. Each author should have made a real contribution to the research and should be involved in writing and reviewing the paper.

You should select the journal to which you will submit your manuscript before you write it or by the time you have a rough draft put together. The best journal for one manuscript is not best for another. Scan titles in the table of contents to be sure you know what subjects are accepted by the journal. Read a few articles closely and examine quality, style, and subject matter. Determine whether the paper will be refereed and who the publisher is. Papers that are refereed, or reviewed, are almost always better because of this process. You want your publication to be in good company. The reputation of the publisher could reflect on your own. Your own society probably publishes quality journals, and some other publishers are equally reputable.

When you have decided which journal to submit your manuscript to, acquire the "Instructions to Authors" from an issue of the journal or by writing to the editor. Many journals publish guidelines annually, perhaps in the January edition, and some have instructions in each issue. When you have studied the journal and its guidelines, write your first draft with the audience and the publication style of the specific journal in mind. If you later decide to submit to another journal, again study the publication and revise your paper to its style.

In selecting the journal to which to submit your work, you should consider circulation and the probable interval between submission and publication. Some journals are distributed worldwide, and the articles may be abstracted by national or international services and data banks. Your particular manuscript may be better for a local, regional, or national publication. Many journals provide information in a footnote on the time from the receipt of the initial manuscript to its publication. This period can be measured at least in months, and often it is well over a year or more from the time the editor receives the manuscript. Some publications are known for a rapid turnaround time, but they may or may not be refereed and known for discrimination in the manuscripts they accept.

Write the paper as you do the research. Get background literature together and write a rough draft of the introduction before your results are available. Write a preliminary abstract without including results. This abstract will help to keep the justification, objectives, and main point in your mind when you begin to consider the results. Write the materials and methods section when you set up the experiment. Then, when results are ready, you can write the results and discussion section(s), the conclusion, and the revised introduction. Finally, revise the abstract by inserting results and a concluding remark. Remember, every section of the paper will have to be revised several times before it is ready for publication.

AFTER THE PAPER IS WRITTEN

When you have written and rewritten your paper and every coauthor has reviewed and revised it, you will reach a point where you can't see how to make the communication any better. It's time then to ask for reviews from your peers. Most institutions or companies you work for and certainly the department in which you are doing graduate work will suggest or require that you obtain in-house reviews before submitting the paper to a journal. Many will require approval before submission and publication. Know the requirements of your department or employer before you submit a paper to a journal.

Whether or not in-house reviews are required, get opinions from colleagues other than the coauthors before you ask for journal reviews. Choose your in-house reviewers carefully. It's good to have two or three. One may be someone who is very familiar with your work. That one may see something important that you had assumed was obvious and failed to include in the text. Choose a second reviewer who knows nothing or very little about what you have been working on but has expertise in similar scientific matters. This reviewer can best give an objective look at both the science and the communication. You may wish to ask for at least a partial review from someone with skills in a special area such as statistical analysis. And a friend, a spouse, or a reviewer who specializes in writing and editing can often improve readability and organization.

Finally, with in-house reviews complete and with further revisions of the paper based on these reviews, you are ready to submit your paper to the journal. Read again the "Instructions to Contributors" or other information that is available with the journal or from the publisher. That information will probably tell you about page charges and submission requirements for page size, line spacing and numbering, and other matters of form and style. Also, consult the checklist for submitting the manuscript in the *CBE Style Manual* (CBE, 1983). It's easy to forget an important point.

Usually the editor will require three or more copies of the paper. Submit the original, except for figures or photographs, and send neat, clear copies. For figures and photographs, make clear copies and retain the originals until your paper has been accepted. You don't want a reviewer marking up your only original. Send them when the editor requests them, usually with a final version of your paper.

Submit your paper to only one journal. Instructions for submission usually indicate that the paper will be considered only if no other journal is considering it simultaneously. You may think your chances for acceptance are better if you try two or three journals, but the publishing staff and reviewers can ill-afford to spend time and money reviewing and editing your paper only to have it published by another journal. Wait for a rejection from the first journal or ask that your paper be released by the first editor before you submit to a second. If you believe the first publisher is taking too long or asking for revisions that you cannot make, ask that the paper be released and send it to another journal; but until then, be patient.

When submitting a paper, include a cover letter requesting that the manuscript be considered for publication in that journal and giving the editor a phone number as well as the address where you can be reached. To indicate the suitability of the manuscript for the journal, you can briefly note what research findings you are reporting. Most editors will send an acknowledgment that the paper has been received and sent to review. If you do not receive such information in about 3 weeks, call the editor to confirm that the manuscript was received. Then wait. If after about 2 or 3 months you have no word from the editor, call again to ask about the status of the paper. It may be lying on an absent-minded reviewer's desk. Your phone call may remind the editor to check with reviewers.

Acceptance of your paper depends on not only how good the research is and how well you write but also on the suitability of the subject and the acceptance rate of the journal. Acceptance can be influenced by the number of submissions with which yours is competing. The rate of acceptance by many journals ranges from 65 to 75% of the manuscripts received, but some prestigious ones accept fewer than 15% and for others the rate may be over 80%.

THE EDITING PROCESS

Editors are human beings. As humans, they have all kinds of personalities; they are sometimes amiable and sometimes surly. Deal with them the way you do other humans; most will appreciate direct, open communication with you. Don't be afraid to discuss and reject editorial changes that could result in a focus or meaning in your paper that you do not want. The paper is yours. What is published is published not under the editor or reviewer's name but under your name.

Editorial staffs are organized in different ways. The journal's Instructions to Authors may outline the review and editing process used. Editors usually log a manuscript in as received. They will then either seek reviews or send it to another editor who will seek reviews and communicate with you. Review processes also differ, but at least two other persons will probably review your paper. On the basis of the reviews and the editor's opinion, your paper will be accepted or rejected.

Seldom will a paper be accepted with no revisions suggested. If this rare event does happen in your life, you will simply be notified that your paper is accepted and going to press. More likely, your paper will be accepted, but revisions will be required. In this situation, your editor will evaluate the reviews, form an opinion, and send recommendations for revisions to you. You certainly may disagree with reviewers, but you should justify in writing to the editor any refusal to accept a recommendation. You'll probably make most of the revisions suggested and send the new copy back to that editor. You may receive a second set of recommendations and have to revise yet again. But when satisfied that the paper is ready for publication, the editor will notify you and put the paper into the publication process.

If the reviewers and your editor believe the paper should be rejected, it will be returned to you with the negative recommendation. The editor will probably provide an explanation for why the paper is not acceptable and may recommend possible revision that would make the paper acceptable. Don't despair when you get that rejection; most of us have experienced the same (Fig. 7-1).

FIGURE 7-1
Don't despair when you get that rejection.

You can certainly discuss the quality or subject matter in the paper with your editor, but don't argue about any recommendation or rejection. That's not professional behavior. If you believe your paper is worthy of publication, revise it again and resubmit it or try another journal.

The following reasons overlap, but one or more of them may explain why your manuscript was unacceptable for publication:

1. The research was inappropriate or poorly conceived and executed.
2. The manuscript was poorly written or did not follow the style of the journal.
3. The results are inconclusive; you have insufficient data or erroneous interpretations.
4. Interpretation is missing or the discussion is unwarranted.
5. The research is trivial (not enough material) or the information is not new or is repetitious of earlier publications.
6. You have too much material; the paper is too long or padded with unimportant data or discussion.

The publication process for professional journals is not perfect. Bad research and bad writing are sometimes published or good research and writing rejected by reputable publishers. However, the process may be as effective as human error will allow. If you will familiarize yourself with what goes on during consideration of your paper, you will be less frustrated with editors and reviewers and with the time required to publish. Reviewers and editors can be very helpful, and your paper will usually be better if you take their advice. If you do not agree with a direct suggestion, you still need to consider it and determine whether you should make some adjustment in the text. Reviewers, who are usually not professional writers and editors, often know something is wrong but cannot identify the exact problem. Take their criticism as an indication that the writing at that point is not clear. Most editors and reviewers are trying to make your paper better, not destroy it, with their suggestions.

Once your paper is accepted, you will have a few final chores to do before you see it in print. You may need to submit a diskette with the final version. Be sure the version on the diskette and the hard copy are exactly the same. You may also be sent a copy of your manuscript called a galley, or galley proof, to be proofread in the form set up for printing. Proofread this galley with great care, and have at least one other person proof it. Finally, you may be asked to sign a sheet to give copyright privileges to the publisher. This signature may transfer your copyright to the publisher, or the publisher may simply be asking for permission to print and reprint your work. Read the fine print to determine whether you retain the copyright. A final chore may be payment of publication charges. Authors are usually charged a fee per page of publication. Your "Notes to Authors" will probably tell you what these charges are.

Always consult a recent issue of the journal to which you submit your work and read the "Notes to Contributors" or other information about that publisher's process. The steps and actions I have described are generalized, and they will continue to change as publishers update their technology to suit their conventions. Many changes are taking place with electronic publication. You may see your publication on an electronic monitor rather than in a bound journal. Lessen your frustrations by keeping up with what's going on.

At least two outstanding books are available for detailed information on writing and publishing a scientific paper for a professional journal. Day's (1994) *How to Write & Publish a Scientific Paper* is a cookbook-style volume that anyone beginning a publishing career in science should have. The other that I find valuable is O'Connor's (1991) book titled *Writing Successfully in Science.* It is not as concise as that by Day, but it is thereby able to include more details. Don't write your first scientific paper for publication without consulting one or both of these valuable sources as well as the journal's notes and a style sheet or manual that a particular publisher uses.

References

Bishop, C. T. (1984). "How to Edit a Scientific Journal." ISI Press, Philadelphia.

Committee on the Conduct of Science (1989). "On Being a Scientist." National Academy of Sciences, National Academy Press, Washington, DC.

Council of Biology Editors (CBE) (1983). "CBE Style Manual," 5th ed. CBE, Bethesda, MD.

Day, R. A. (1994). "How to Write & Publish a Scientific Paper," 4th ed. Oryx Press, Phoenix, AZ.

O'Connor, M. (1991). "Writing Successfully in Science." Harper Collins *Academic,* London.

8
STYLE AND ACCURACY IN THE FINAL DRAFT

"The simplest rule of thumb for the author and illustrator should be: follow what you see in print."

COUNCIL OF BIOLOGY EDITORS

Attention in early drafts of a paper should focus on organization and content, but you should also be conscious of format and details that are essential to the final draft. Before the paper goes to reviewers, you need to check carefully for accuracy and form in style and documentation. The final chore is to proofread carefully so that reviewers are not distracted by errors in style and grammar.

STYLE

In scientific writing when we refer to style, we are usually not referring to the effect that your personality or literary training has on your writing but rather to technicalities and details in the form that your work takes. This **technical style** includes what is acceptable or unacceptable in organization, print size and style, capitalization, punctuation, indentions and spacing, abbreviations, citations and bibliographies, headings, footnotes, and any other conventions in the format or content of the entire work. Style is the way publishers or editors want to print your text to give a physical as well as verbal appearance to their publications and to your paper. The way words are physically arranged, capitalized, and punctuated can add clarity to the communication.

Unfortunately, no consistent standards for style exist in the scientific community, sometimes not even within individual professional societies. Study carefully the style of the publisher to whom you intend to submit your work. Almost all follow similar organizational patterns, but the headings will differ. What one calls Methods, another may call Procedures; and where one may want you to combine the Results and Discussion sections, another may separate the two. Notice also the position and type styles used in headings and small details like abbreviations and punctuation. An editor or reviewer is much more likely to read your paper with a positive attitude if you have been attentive to style.

Although styles differ widely, some efforts toward uniformity have been made in recent years. The international system for units of measure is an important step forward. Biomedical disciplines have made an attempt to adopt uniform requirements in manuscripts submitted to their journals, even if the publisher edits to an individualized style (O'Connor, 1991). The Council of Biology Editors (CBE, 1994) has produced *Scientific Style and Format* with a goal of providing uniformity in style for all the sciences. Other style manuals (e.g., CBE, 1983; Dodd, 1986) provide more discussion on issues such as planning, writing, and submitting the paper for publication, but they are less comprehensive on stylistic details. Most large professional societies print still more limited style manuals for their contributors (e.g., American Society of Agronomy, U.S. Geological Survey, U.S. Government Printing Office). A list of style manuals is provided in *Scientific Style and Format* (CBE, 1994). In addition, each scientific journal publishes its own style sheet titled "Notes to Authors," "Suggestions to Contributors," or a similar title.

Decide where you wish to publish early in your writing and carefully follow the style sheet for that publisher. Failure to do so may mean that an editor will reject the manuscript even before he or she sends it to review. The title page near the front of the journal usually contains documentation about the publication and includes information about such things as subscription rates and publication charges as well as where to find information on the journal style.

STYLE IN HEADINGS

Manuscripts for journal articles, chapters in books and theses, proposals, or other documents or publications often need to be subdivided into sections. Headings provide transitional guideposts to link these divisions. Check a style sheet or an example of a publication to determine spacing as well as heading and subheading patterns, or make your own style neat, consistent, and attractive. Most important is to be consistent in print style, size, and position for headings that are parallel.

Headings are usually designated as primary, secondary, and tertiary heads or first-, second-, and third-level headings. Unless your text is book length, you will probably not need fourth- or fifth-level headings. If a style sheet does not require that you follow specific patterns for headings, you may wish to use some of the following. Notice the symbolic emphasis given to the various forms. Positioning, spacing, boldface, underlining, italics, capitalization, or other symbolic dimensions can be manipulated to provide more or less emphasis to a heading. Determine which of these your publisher uses, and be consistent with parallel headings. I have attempted to arrange the following headings relative to their importance, but various forms of enhancement can change this order.

Sample headings

1. Center
   A. Underline and/or boldface

      Results and Discussion

   B. Use total capitals

      RESULTS AND DISCUSSION or **RESULTS and DISCUSSION**

   C. With no added enhancement

      Nitrogen Study, 1996

2. Left margin
   A. Above line, underlined, italic, or boldface

Nitrogen Study, 1996

Text begins here . . .

   B. Above line, not embellished

Nitrogen Study, 1996:

Text begins here . . .

   C. On line, italic, or boldface

Nitrogen Study, 1996. Text begins here . . .

   D. On line, paragraph indention, underlined, italic, or boldface

      Nitrogen Study, 1996. Text begins here . . .

   E. Pattern C or D with no embellishment

ACCURACY AND STYLE IN DOCUMENTATION

A most obvious point at which styles still differ widely among journals is in their techniques for handling the references. Both the form for citing the reference in the text as well as the form used in the bibliography vary from one journal to the next. Even the heading for the bibliography differs. Most of them are called

"Literature Cited" or "References." The two terms have slightly different meanings, but both are bibliographies limited to works cited in a text. A *bibliography* is a list of works related to a text. These works may or may not be literature (i.e., published) and may or may not be cited in the text. *Literature cited* usually refers only to published literature that is cited in the text. *References* may or may not be published forms but are cited in the text. Not all publishers adhere strictly to these definitions. Disagreement also exists about what is or is not considered literature or published.

The most essential criterion for documentation is that it be accurate and complete so that your reader can readily find any source of information that you have referred to. Every citation must be documented clearly and completely according to the style of your publisher. When a style for citations in the text and in the bibliographic list is not dictated by your publisher, at least be consistent and accurate. If you give erroneous details about a source, you'll mislead and frustrate another researcher. Failure to give accurate or adequate information can reflect on your credibility.

Details differ in the reference lists for the various journals. Punctuation, capitalization, use of first names or initials and their placement—it's really amazing how many ways publishers can find to alter styles. It is easier to keep up with references if you use the name–year system as you write the paper even if you have to change to the number system or other styles when you write final drafts. The following are samples of some basic styles for references, but don't trust them entirely. Look to the journal to which you are submitting your work; its style is sure to differ somewhat from any of my examples.

Three systems for textual citations and bibliographies:

1. ***The number system*: Alphabet–number system**

Examples in the text

A. In 1968 Bilbrey and Rawls (2) developed a technique for. . . .
B. With the mathematical model (2), we could project. . . .
C. Several theories have been proposed for measuring soil water potential (2, 4, 7, 13, 21).

Advantages and disadvantages

This system provides little interruption to the text, is less costly to the publisher than is the Harvard system, and can feature the name of the author(s) or a date in the text as in example A. However, the number system does not always give immediate identification by name and year. If the writer adds or deletes a reference, he or she must carefully renumber citations throughout the text as well as in the bibliography.

The list of references is alphabetized and numbers correspond with the citations in the text.

Sample bibliography for the number system

Literature Cited

1. Adcock, R. L. 1988. Effect of moisture stress on soybean pod development. Crop J. 95: 345–347
2. Bilbrey, J. C. and R. M. Rawls. 1968. Measuring soil water potential in a Sharkey clay. Gen. Soil Sci. J. 13: 121–124.
3. Green, C. R., A. C. Dobbins, V. C. Martin, and W. R. Amity. 1986. Response of grapes (*Vitis lubrusca*) to drip irrigation. HortReport 59: 13–14.

2. *Name–year system*: Harvard system

Examples in the text:

A. Bilbrey and Rawls (1968) developed a technique for. . . .

B. By using a mathematical model (Bilbrey and Rawls, 1968), we could project. . . .

C. Several theories have been proposed for measuring soil water potential (Adcock, 1988; Bilbrey and Rawls, 1968ab; Dobbins, 1981; Ferguson and Fox, 1979; Fox, 1991; Lennon et al., 1992; Watson et al.,1985). These theories allow for. . . .

Advantages and disadvantages

With this system the writer can add or delete references easily. Some immediate identification of the reference is apparent. However, several such citations in a row can distract, and publishing cost is greater than that for the number system.

The list of references is always alphabetized and may be numbered as well.

Sample bibliography for the Harvard system

References

Adcock, R. L. Effect of moisture stress on soybean pod development. Crop J. 95: 342–345; 1988.

Bilbrey, J. C.; Rawls, R. M. Measuring soil water potential in a Sharkey clay. Gen. Soil Sci. J. 13: 117–121; 1968a.

Bilbrey, J. C.; Rawls, R. M. Drip irrigation for grapes. HortReport 78:14–15; 1968b.

3. *Order of citation system*: Vancouver system

Textual citations and bibliographic list are numbered in the order in which the citations occur in the text.

Example in the text

In 1975, Fox[1] developed a technique to measure soil water potential. Bilbrey and Rawls[2] modified that technique in 1980 to the form used today[3,4,5].

Advantages and disadvantages

This system is simple for printing short papers with short lists of references. A specific reference can be difficult to locate if the list is long.

The list is not alphabetized but is numbered and listed in the order of occurrence in the text.

Sample bibliography for the order of citation system

References

1. Fox, R. T. (1975) Agric. Bull. 102,47–49.
2. Bilbrey, J. C. and Rawls, R. M. (1980) HortReport 32, 17–21.
3. Lennon, T. R., Elzie, M. S., and Cola, R. C. (1992) Crop J. 79, 173–177.

In the preceding samples, various styles are used. These forms are all acceptable in some publications, but none may be appropriate to the editor and publisher to whom you submit your work. Always carefully follow the style of the publisher for whom you are writing. (All the examples shown here are fictitious.)

DOCUMENTATION OF ELECTRONIC SOURCES

With more and more information being acquired from computer networks, clear citation to those sources is equally as important as reference to printed material. The important point in documenting material is that a reader can find the same source with the same information you retrieved. But standards for storing and documenting electronic sources are not yet stable. The material on-line is not archived as is printed text, and it can be changed or even removed from the network. It may be important to list the date you accessed the information as well as the date it was entered on-line. Problems in documentation of electronic sources are far from solved. The CBE style manual (CBE, 1994) has some basic information on formats for electronic documentation as recommended by the National Library of Medicine. If you need to use such citations, your editor may also have suggestions. Be alert to new information on electronic documentation.

Headings and documentation are just two prominent areas in which styles differ markedly between publishers. You need to be conscious of other elements of style as well. Use of abbreviations, spacing, footnotes, and other inclusions will differ. With the capabilities of word processing and of electronic publishing, more and more of the responsibility for style will come to rest with the authors of a paper. Certainly you need to be conscious of the communicative value of how a page looks to a reader as well as what the words say. Above all, be sure that you have proofread your final draft so that you don't burden the reader with all the distracting errors that slip by during composition. As C. C. Colton says, "That writer does the most, who gives his reader the *most* knowledge, and takes from him the *least* time."

PROOFREADING

Probably nothing blows smoke in the eyes of communication more quickly than the small mechanical error. Don't try to make early drafts of a paper perfect, but when you reach the final stages before sending a manuscript to your reviewers, then proofread as though the entire acceptance of the paper were at stake, and it may well be. Readers come to your work from outside. They may pay attention to content until they stumble across the misspelled word, the transposed letters, the word out of place. Then the mind jumps to that small point and can entirely leave the content. The reader's mind is ticking with: "That's not right. Isn't it *i* before *e*? Shouldn't that be a capital? I never could spell that word either . . . " and so on until what you were saying gets lost in something like two small letters transposed.

Often people let important decisions about your reputation hang on the grammar and mechanics in your writing. "Look there. A misspelled word in the first line. Don't they teach spelling any more?" Before those readers are finished with reading, you are labeled a careless, uneducated individual, and they won't hire you nor publish your work. Perhaps I exaggerate the importance of accuracy, but I think not. Learn to proofread.

Proofreading is not editing. The proofreader is looking for the mechanical or grammatical error and only indirectly at the content. Proofreading is not like any other kind of reading. There is scanning, there is careful reading, there is studying texts for content, and there is speed reading. But none of those are proofreading. If I am asked to proofread a paper I am not familiar with, I proofread it at least three times. First, I read through the paper so that the basic content will not distract me the second time through. During that first reading, I catch several errors that I check lightly. Then I read the text word by word, phrase by phrase, and scrutinize the innards of the words, the order of the phrases. As I find errors, I mark them lightly again. The third time I read with the flow of the words, sentence by sentence, paragraph by paragraph, yet again noting any error I missed in earlier readings. Then I go back and mark each error carefully with the appropriate proofreader's symbol. I am likely to read the text aloud a fourth time after it is marked to verify my marking and catch that error I missed.

With word processing some proofreading can be done on the monitor and corrections can be made easily. Some people read from the screen more easily than others. I recommend that at least one reading be done from hard copy. The screen may not tell you exactly how your work will look on the printed page. Use the spell check feature in word processing, but don't trust it implicitly. It misses real words that are used incorrectly.

If you have trouble catching errors, read aloud. Or have someone read slowly to you while you follow the copy. Always have a second reader proof your final draft. It is most difficult to catch the errors in your own work. You know what you meant to say, and your mind will read in omissions or leave out any element that intrudes. Proofread and mark copy carefully. The following are some handy symbols to know. Others are listed in unabridged dictionaries or style manuals.

Common Proofreader's Marks

Correction	Marginal Symbol	Marked text	Corrected text
Delete	ℊ	the ~~old~~ cat	the cat
Restore deletion	stet	the ~~old~~ cat	the old cat
Close up space	⁀‿	the o ld cat	the old cat
Insert	old	the^cat	the old cat
Replace	old	the ~~big~~ cat	the old cat
Insert space	#	the/cat	the cat
New paragraph	¶	^Once upon a time	Once upon a time
Move to left	[	[the dog	the dog
Move to right	]	the dog]	the dog
Center	ctr	]the dog[	the dog
Transpose	Tr	the on dog	on the dog
make lower case	lc	the ~~O~~ld dog	the old dog
Capitalize	caps	jim burns	Jim Burns
Period	⊙	Go west^	Go west.
Comma	^,	Come^Jim.	Come, Jim.
Apostrophe	’	Jims dog	Jim's dog
Superscript	^{2}	3 m2	3 m^2
Subscript	$_2$	H2O	H_2O

References

Council of Biology Editors (CBE) (1994). "Scientific Style and Format: The CBE Manual for Authors, Editors, and Publishers," 6th ed. Cambridge Univ. Press, Cambridge, UK.

Council of Biology Editors (CBE) (1983). "CBE Style Manual," 5th ed. CBE, Bethesda, MD.

Dodd, J. S. (Ed.) (1986). "The ACS Style Guide: A Manual for Authors and Editors." American Chemical Society, Washington, DC.

O'Connor, M. (1991). "Writing Successfully in Science." Harper Collins *Academic,* London.

9
REVIEWING AND REVISING

"No one can make you feel inferior
without your permission"

ELEANOR ROOSEVELT

For almost any craft or art form, the initial rough creation has to be reshaped, sanded, revised, polished, or fine-tuned as do the draft and subsequent forms of a scientific paper (Fig. 9-1). In other words, good scientific communication is not the product of an initial attempt by a talented writer but rather the end result of repeated reviewing and revising. Revising is essential, and good reviews can be helpful in this process. Few people write well; many people revise well.

Recognize that a well-written paper will be reviewed and revised at least three to six times before it is acceptable to your audience. Many good papers are revised far more than six times before publication. A barrier to achieving a well-done final draft is a failure to admit that you, like everyone else, need the opinions of others and need to revise more than once. I assure you that no chapter in this book has had less than six revisions, most of them far more, and that every review has been helpful. As a scientist, you will review and revise your own work, use reviews from others in revising your work, and serve as a reviewer for other authors.

FIGURE 9-1
Good communication has to be polished from an initial rough draft.

REVIEWING AND REVISING YOUR OWN PAPER

Learning to review and revise your own work requires an attitude of confidence but of objectivity and humility as well. To look at your own work objectively is difficult, but with an open mind, you can develop an ability to do so.

After you have written your first draft, carry your own reviews through at least three stages with revision. You will first review and revise the general content to determine that you are emphasizing the main point of the paper. Then, revise important parts of the paper to support the main point in the methods, the results, the supporting literature, and the tables and figures. Finally, pay attention to clarity and details in style, diction, and mechanics. Try some of the following ideas.

Lay your work aside for several days or even 2 or 3 weeks. If deadlines are pursuing you, at least take a break for a few hours. The psychological adjustment you can make by getting away from your own writing may save you time in the long run. When you pick up the paper again, try to read it as though you were completely unfamiliar with the research or the writing. Read through the paper without stopping to criticize small points just as though you were reading a published article. With the overall picture in mind, ask yourself whether "this author" is emphasizing the main point, whether he or she has organized the material well, and whether too much or too little is said. In other words, adopt an outsider's attitude and criticize the general content. What questions would you have if you were not familiar with the work? Try to recognize whether the hypothesis, the objectives, the methods, and the results are clear to another scientist. If not, begin work on those points without worrying over details of sentence structure and grammar. At this point you are looking for general clarity, logic, and a focus on your point of emphasis.

When you have revised and are convinced that your main points are clear, read the revised version—this time much more slowly. Check content in the main sections of the paper: The abstract, the introduction, and so forth. Check through the materials and methods section carefully. Could an outsider follow your instructions and produce the same experiment with the same results? Check your own data for accuracy. Look at the tables and figures and the text of the results and discussion. Can you simplify data for the outsider; do you need to add a point; or more likely, can you delete some discussion or even a column in a table or a whole figure without diluting a point? Check ideas in textual citations and in the references; consult the sources themselves for what they say. Have you distorted a citation with the context in which you refer to it? You are not going to do this sort of thing deliberately, of course, but it is easy to misinterpret an author or make an error in transferring details or ideas to your own words.

After you have revised the content of your paper in this way, read it again very slowly, phrase by phrase. This time watch the diction and internal organization. Watch for smaller flaws in logic and structure of the sentences. Does each statement support your hypothesis and your line of reasoning? Remember Day (1994) and Tichy (1988) and Zinsser (1988). Are your sentences full of clutter? Rewrite them. Watch for precision and details. Check accuracy in spellings, numbers, meanings of words; adhere to journal style. Don't trust your good memory or your intuition. Check data against original records. Especially verify all textual citations and references.

After you have gone through these three stages in the first revision, read the paper through again smoothly. **Read it aloud.** It's amazing what you hear that you don't catch when you read silently. Is there something you missed in the revisions? You may need to repeat all three stages again, or the last two, or the last one. Most people would do well to put the paper aside again for several days and then go through the revision process step by step at least once more. At some point along the way, however, you will do well to turn the criticism over to coauthors or reviewers. Revise again after you receive their comments.

MAKING USE OF REVIEWERS' SUGGESTIONS

Before you send your work into the official review process, be sure you have done your own and in-house reviews.Then, don't expect miracles from the publisher's reviewers. Although peer reviewers are helpful, they will not rewrite your paper for you or make a good paper out of a bad one. Send to them a version that you believe you yourself cannot improve. With most professional journals, the reviewers are volunteering their time. The only compensation they get is the satisfaction of participating in professional responsibilities. Allow them time to do a good job. Treat the reviewers with respect by first getting the paper to the point where you think no further revisions are needed and then by giving serious attention to their critical observations.

When you review a paper for someone else, you should be candid with the author about what you see as strengths and weaknesses in the paper, and you should appreciate the same kind of candor for your own manuscripts. The reviewer has a perspective that is closer to the audience than the author's is. Take seriously the review comments and use them to your advantage. You don't have to agree with your reviewers, but you need to respect their opinions.

Again, your attitude is important. Be prepared to revise yet again. Even though you submitted the paper fully believing that it was in need of no further revision, you will need to revise after the reviews. I have worked with hundreds of refereed papers. **One** among them was accepted for publication by a refereed journal with no revision requested. That one was a methods paper by a graduate student. Don't send your paper away expecting to win such a lottery. The chances are remote.

If you expect to receive numerous suggestions for revision, you will be ready for what you receive. Read the reviews objectively. Generally, the reviewers are sincerely interested in helping to produce a good publication and are not trying to find fault with your paper. Approach all the reviews with this idea in mind, and try seeing your paper from their point of view. Your goal is to produce a better paper after incorporating or otherwise acting on their suggestions.

Bad reviews can happen to good papers, and as Danielle Steel says: "A bad review is like baking a cake with all the best ingredients and having someone sit on it." Probably the worst reviews are the nonreviews or those that say very little. General comments such as "Tighten the organization" or "Well written, could add more data" are not very helpful. The reviewer needs to suggest points at which the organization is weak or where you need more supporting data. Without rewriting the paper, reviewers can and should be specific.

Accept the fact that occasionally you will encounter reviewers who consider that only they understand your subject. They may play "king of the mountain" to try to push your research aside so that theirs remains on top. Don't let such people frustrate you. They can really be unintentionally helpful. In their criticism, they will likely allude to the weakest parts of your paper, and even though you may not agree with their criticism, you can make the effort to strengthen that part of the manuscript. Any reviewer who finds something wrong, weak, or hard to comprehend is doing you and your paper a service no matter what his or her attitude is. The reviewer may not be able to pin down what is wrong with a part of your paper, but the fact that he or she is questioning it probably means it could be improved with rewriting. Maintain your own confidence in what you are doing, and let the criticism lead you to a better paper.

Be open-minded but do not blindly follow suggestions by reviewers. They may be entirely wrong about a point. You may need to explain to an editor why you have not followed a reviewer's suggestion, but remember that the paper belongs to you and any coauthors who are working with you. Reviewers are people too; they make mistakes. Just don't conclude that the mistake is theirs until you have engaged in some objective self-criticism. Then with an open mind revise your paper yet once again. Lay it aside a few days, and then review it yourself. When you believe the accuracy and clarity of the paper are pristine, return it to the editor. If each author, reviewer, and editor in the publication process performs his or her responsibilities in a professional manner, we can be proud of the scientific literature, including your paper.

REVIEWING THE WORK OF OTHERS

The most common encounter you will have with reviewing for others will probably be with manuscripts submitted to professional journals. Journal reviews can be full-blind reviews in which the authors' names are not revealed to the reviewer nor the reviewer's name to the author. At the other extreme, reviews are occasionally fully open; the author and the reviewer may not only be identified to each other, but along with the article, the journal may also publish the reviewer's comments and even a reply from the author. Many reviews fall somewhere between these two extremes; often the author does not know the reviewer, but the reviewer knows who the author is.

When you review a paper for someone else, three professional principles should be utmost in your mind. *One, do a good job.* Along with editors, reviewers are the guardians of the scientific literature; enough garbage slips into publication without your contributing to it by not giving due attention to a review. *Two, review in a timely manner.* Authors are anxious about their papers. Get your review back to the editor in the time requested. *Three, keep information in the manuscript confidential.* The work is the property of the author. You may run across exciting ideas in your reviews, but they are not yours, and public disclosure or your own use of the information is unethical until the work is published and then clearly referenced.

When you review a paper for someone else, you can follow the same plan that I've outlined for reviewing your own paper. This time, however, you don't have the intervening revisions. You simply review and comment on major points, sections, and details from the same version. Keep in mind that your job as reviewer is not to revise but simply to make suggestions and let the authors act on those suggestions as they see fit.

The peer review process for journal publication is certainly not foolproof, but it may be our best safeguard against too much inferior literature in the sciences. As a reviewer, reception of the paper by the audience is your chief concern, but you are in a position somewhere between the editor and the author. Try to understand whether your position is midway or skewed in one direction or the other. Sometimes the reviewer's major responsibility is to the author of the work to help get the manuscript ready for an editor; sometimes the major responsibility is to the editor to help judge and justify the acceptance or rejection of the paper and make specific recommendations to the author. Whatever your position, you should carefully *evaluate the work to the benefit of both the editor and the author on behalf of the audience.* Keep in mind that yours are mere suggestions. Both editing and rewriting are outside the realm of your responsibilities as a reviewer.

The reviewer's role is to peruse the entire contents with full attention to all details and accuracy in the communication. To be a full reviewer, you should be knowledgeable about the subject of the manuscript, you should have read much of the literature on the subject, and you should be familiar with the journal to which the manuscript is submitted and the audience who will be reading the published work. Only with this expertise can you judge overall quality in accuracy of scientific details and clarity in communication. Because the reviewer's role is one of assistance, you should maintain the attitude of humble but confident objectivity. The reviewer must express an opinion on the overall values of the manuscript and strengths and weaknesses in its component parts. The advice to reviewers in the *CBE Style Manual* (CBE, 1983) can be helpful to you in establishing your role and in developing an effective attitude. The following suggestions may also prove useful.

Keep your own paper handy, or make a photocopy of the manuscript to work with. Initially at least, do not write on the manuscript itself. You can well change your mind about your own opinions as you study the manuscript, or the editor or author may have requested that you not mark the copy.

Read through the entire manuscript to gain a familiarity with it. Then read it carefully and consider the following questions: Is the research clearly justified and made credible with appropriate allusions to scientific principles and the literature? Does the work add to the knowledge already available in a given area of study or is it essentially a duplication of an earlier publication?

If you can give positive replies to those considerations, consider the research itself as depicted in the paper: Is it valid, complete, and credible? Are the hypothesis and objectives clear? Is the statistical design appropriate, and are the methods presented in adequate detail for a reader to repeat the experimentation? Are the results clearly stated, and do the data support the interpretation given to the results? Are the data analyzed statistically, interpreted accurately, and presented in text, figures, or tables that are easily comprehensible?

Now read it carefully again and consider the manuscript by parts and in detail. Are the main points emphasized in the right way? Is the title appropriate? Is the abstract the right length, and does it contain the adequate details? Should some sections of the manuscript be deleted, revised, shortened, or expanded?

Then, as in reviewing your own work, go to the smaller details. Is organization within paragraphs clear, and is the sentence construction effective? Does the author follow the proper format for the text, the tables, the figures, and the references? Are all important references cited, and are any of them simply excessive baggage? You may even mark misspellings or other mechanical and grammatical problems that you see in the writing. But remember that your role as a reviewer is not to be a technical editor and that you should not rewrite for the author nor edit for the editor in your role as a reviewer.

You are making notes on all these matters or marking a work copy as you read and reread the paper. Finally, prepare your comments for the editor or author on the manuscript itself or write a critique on a separate sheet. Write your comments clearly with specific suggestions for how to make improvements. It doesn't mean much to an author simply to say, "Not acceptable." *What* is not acceptable? Do you have a suggestion for making it acceptable? In other words, be specific without actually rewriting material.

Finally, ask yourself whether you can support your criticisms. Have you been helpful or prejudiced? And have you communicated your opinions clearly to the editor and the author? Look at the review presented in Appendix 5. A good reviewer is invaluable; a bad one is demoralizing and destructive.

References

Council of Biology Editors (CBE) (1983). "CBE Style Manual," 5th ed. CBE, Bethesda, MD.

Day, R. A. (1994). "How to Write & Publish a Scientific Paper," 4th ed. Oryx Press, Phoenix, AZ.

Tichy, H. J. (1988). "Effective Writing for Engineers, Managers, Scientists," 2nd ed. Wiley, New York.

Zinsser, W. (1988). "On Writing Well," 3rd ed. Harper & Row, New York.

10

TITLES AND ABSTRACTS

"All words are pegs to hang ideas on."

HENRY WARD BEECHER

Titles and abstracts are the parts of your paper that will be read most often. They serve two purposes for your readers: (1) to disclose the basic information that the paper itself contains and (2) to help readers decide whether to read the entire paper. Together the title and the abstract must help the reader quickly survey the literature. As you write them, keep these ideas in mind.

TITLES

The title is the first impression you make on your audience. It should attract attention, but most important, it should be informative. The title may be the most notable phrase you write. Many people will read it, but few will read the rest of your paper. It should use

1. **The most precise words possible**
2. **Words that indicate the main point of the paper**
3. **Words that lend themselves to indexing the subject**

One technique for creating a title is to write the objectives first; then write the rough title, sometimes called the *working title.* Go on to write the entire paper, and then revise the title. Write and revise the abstract, and then check the title again. It may need another revision.

The journal may also request a *running title*, or *running head*. This is simply an abbreviated form of the title that appears as a headnote on journal pages beyond the first page of an article. For more detailed information on titles, check Day (1994). No, let me be less casual. **Read** Day on titles; his remarks will give you a feel for what to do with your own title.

The most common problems with titles perhaps are in their length and in the selection and arrangement of words. Be sure your title will make sense to someone not familiar with your subject. Use words that other readers might consult to find information such as your paper contains and use them in a sequence that is not ambiguous or misleading. The first source for key words for indexing services is the title. Study your title for unnecessary words and put the most important ones first. Provide adequate information but don't make your title too long. Eight to 12 words is a good range to work in. Don't try for a headline. Scientific readers aren't looking for a journalistic sensation story; they want information. A full sentence with an active verb is usually not a good title. Just be informative. Take a look at the various versions of the title in Appendix 6.

ABSTRACTS

The term *abstract* is used loosely to refer to almost any brief account of a longer paper. Informative abstracts used with scientific journal articles are a more structured form than this loose definition permits. With reference to *BIOSIS Guide to Abstracts* and to the *American National Standard for Writing Abstracts*, the *CBE Style Manual* (CBE, 1983) distinguishes between the other brief forms and abstracts that are used with journal publications. What that manual describes as an abstract is called an informative abstract by Day (1994) and others who compare it to a descriptive abstract. To complicate definitions further, some societies will request "extended abstracts" for publication in proceedings. The best I can tell, these are much shorter than a full paper but can contain more data than the informative abstract because the publisher is providing more space for the publication. Observe a previously published issue of proceedings to determine what is expected if your society requests an extended abstract.

The **descriptive abstract,** or topical abstract, describes the contents of a paper but does not give a full condensation of the information contained therein. Its contents would be relatively worthless if it were not accompanied by the report itself. It may be the best form for some reports and, like a table of contents, is helpful for a reader in deciding whether to read the entire paper. But descriptive abstracts contain too little substance and detail to substitute for the informative abstract.

Don't let all this fuss over definitions misguide you. Just know that these strange breeds exist, and then recognize that for journal publications you need the **informative abstract.** Study the journal to which you will submit a manuscript to obtain specific instructions. You will find a few differences between journals, but any informative abstract must serve several purposes:

1. **To show the reader very quickly whether the full report is valuable for further study**
2. **To be extracted (abstracted) from the full report for separate publication**
3. **To furnish terminology to help in literature searches by individuals or by literature retrieval specialists for indexes and computer banks**

To serve these purposes, the abstract must be a short, concise, but completely self-explanatory report on a scientific investigation. Like the report itself, the abstract must include

1. **The research objectives and basic justification for conducting the investigation**
2. **The basic methods used**
3. **The results and significant conclusions that can be drawn**

Notice that the two parts generally included in the full paper that are omitted from the abstract are the literature review and discussion. A concluding statement may give an interpretation to the results, but any lengthy discussion or speculation is out of place. Most style sheets specify that the abstract should not exceed 200 to 250 words or 3 to 5% of the length of the paper itself and that the form should be one paragraph. Even for less formal presentations, such as papers presented at meetings or published in nontechnical forms, the abstract should still be a technical, concise, complete report. If such precision is out of place, an informal annotation or other summary should be substituted.

Societies sometimes publish lists of what should be included in or omitted from an abstract. Let me try to hit upon some main points and refer you to the *CBE Style Manual* (CBE, 1983) and Day (1994) for what the abstract should include. Along with the essentials (i.e., research objectives, basic methods, and results), the abstract should fully summarize the contents of the paper in as much detail and as few words as possible. In being concise, don't sacrifice clarity for a telegraphic style that is hard to read. Because it is often read hurriedly in scanning the literature, it should flow smoothly. Emphasize the main points and avoid long lists of information. In presenting the main point, be as specific as possible. For example, say "20 and 40 kg ha^{1} of nitrogen" and not just "two rates of nitrogen." Keep the tone strictly objective; don't use any loaded language that would even suggest speculation. Provide any scientific information, such as scientific names for species, that is important for a complete understanding of your subject. Avoid abbreviations that are not immediately evident to the scientific community. Avoid allusions to the literature or to any other material that would require a footnote or pursuit of external information.

When you think about it, all these do's and don't's for the abstract simply point up its purpose: To be a concise, complete report of your work that can stand alone without further explanation. Notice how the abstract in Appendix 7 evolves from a wordy working abstract to a more concise version for publication.

References

Council of Biology Editors (CBE) (1983). "CBE Style Manual," 5th ed. CBE, Bethesda, MD.
Day, R. A. (1994). "How to Write & Publish a Scientific Paper," 4th ed. Oryx Press, Phoenix, AZ.

11
PRESENTING DATA

"Graphic excellence is that which gives to the viewer the greatest number of ideas in the shortest time with the least ink in the smallest place."

EDWARD R. TUFTE

Once you have analyzed the data from a research project, you are ready to put them into a form to communicate to others. Before you decide what form your data will take, consider your purpose in presenting the data and the audience for which it is intended. Ask yourself what point you wish to make. Also, be sure that you yourself understand the point, the data, and the data analysis and firmly believe in what you are reporting. Then deliver the information with as much clarity as possible for the form of communication you are using—a journal article, a poster, a slide presentation, or other report.

You may be able to present data in the flow of a text. But often tables, graphs, maps, photographs, flowcharts, or other figures or illustrations can communicate more clearly than text can. It is not realistic to try to use all the data points that you collect during your experimentation. As Gastel (1983) suggests, "Good science communication, like good science, usually entails gathering much more information than will appear in print." Your job as a writer or speaker is to select representative data and to determine which method of presentation will be most clear for the audience. Some audiences need more verbal explanation and might not understand linear regressions or logarithmic scales; others might be offended if you do not use these forms.

Fortunately, we have a wide selection of formats to use for data presentation. Once you have the audience in mind, think in terms of which format will best express your point. **Tables** are excellent for presenting specific data and making exact comparisons between data points. Tables can also show gradations and relationships between controls and treatments. **Bar charts** are not as numerically specific as tables but can make more dramatic comparisons. Bar charts make comparisons in sizes, magnitudes, amounts, and other distinctions. They should emphasize differences rather than trends, but as with data in tables, the bars can be arranged to indicate a trend. Although measurements can be closely estimated relative to a scale on the axis, the purpose of a bar chart is not to show specific numbers, and seldom should you duplicate data by writing specific amounts above the bars. If you want the specific numbers, use a table. **Line graphs** are designed to demonstrate movement, change, and trends, especially over time or concentrations. Like bar charts, their purpose is not to show specific amounts, but they can show estimates against a scale. The semilogarithmic line graph can demonstrate relative changes in two values. Variations in tables, bar charts, and line graphs give you a wide range of choices. Select the form that will make the data say what you want the audience to understand.

Results in a report are relatively easy to write or present once you have displayed your data in tables and figures and studied their meanings. Keep in mind the communication principle of simplicity. The simplest possible message is the most likely to be understood. Seldom, if ever, do you need to use both a table and a figure to express the same information in written work. For a slide presentation, you may want to use both in order to move from one slide to another and give the audience two views of an important point while you talk about it.

In any communication, misunderstanding is a danger. It is imperative that you present your results with every degree of clarity and honesty possible. Of course, deliberately loading data to make a point is a cardinal sin, but even inadvertently giving the audience a false impression by inaccurate or sloppy presentation of data or a failure to adhere to conventional use of words, tables, and figures is as bad as guessing at measurements in the laboratory. Precision, accuracy, and honesty must continue throughout data analysis and presentation.

The ease with which data can be analyzed by computer and the comparable ease with which you can devise a table or a graph with the computer are invaluable advancements for scientific communications. However, any technology involves both beauty and danger. The computer should not dictate the kind of analysis and presentation you will use. Recognize the conventions used in tables and illustrations of all kinds. The readers will interpret data according to conventional techniques or by their visual perception of what is most important. Cleveland and McGill (1985) and Tufte (1983) describe some of the pitfalls that occur with presentation of data in figures. These misconceptions result from visual or psychological illusions because of the arrangement of data relative to simple communication devices involving color, intensity, size, and spacing. Any amount of explanation will not completely overcome the psychological impact of a wide line compared to a narrow one or a large letter compared to a small one. Size suggests importance. The same would be true of a column of information in a table if it were set apart in some way with color or spacing. Guard against misleading an audience with poorly presented data.

With and without statistical analyses, your data can be complex. Your job is to simplify your message without falsifying the data. Every scientific experiment has exclusive variables or details that need to be emphasized, and results may require unique methods of presenting data. Because of the many differences in portrayal of data, you should study the journals published in your scientific discipline. As the Science Illustration Committee (CBE, 1988) concludes: "The simplest rule of thumb for the author and illustrator should be: follow what you see in print." The following observations may be helpful.

TABLES

If a table is used, don't repeat exactly what it says in the text—just call attention to its main points. A table should be able to stand alone, i.e., to communicate a point or points without the need to refer to the text. Background material such as information on how an experiment was conducted will be given in the text, but the reader should be able to interpret data presented in a table without referring to the text. Definitions needed for understanding can be placed in footnotes or captions and headnotes. Abbreviations that cannot be used in the text are sometimes appropriate for tables, but meanings not immediately clear should be defined with the table. For clarity and understanding, a table may repeat information in more than one form—i.e., in addition to the absolute values presented, it may give percentages, totals, means, averages, or ratios of those values.

Authors are more likely to put too much than too little information in a table. Limit the items in the field, especially in tables used for slide presentations. Often full columns of information can be omitted. Be suspicious of any column that has few or no data points that differ. A string of 0's or a column that repeatedly notes the same result can often be omitted. A full table consisting of no more than six or eight data points can usually be replaced by a sentence or two in the text. Tables are also overloaded when headings and footnotes are too heavy. If an abbreviation is obvious from information in the caption, that abbreviation can be used with no further explanation.

Problems that occur with tables may be crucial to a reader's understanding. Errors can occur easily in transferring data to a table or in subsequent revisions. Captions, like titles, are sometimes too long and wordy; they should be concise and contain only the key words needed to clarify the message in the field of data. Numbers should show no more decimal places than are essential for reasonable accuracy and understanding. As with other communication, try to look at your tables from the viewer's perspective. Make them precise, concise, simple, and clear. The following briefly illustrates parts of a table and the terminology used to name those parts:

Table No. Caption or title

	Main boxhead (identify items in field)[a]		
	Secondary head No. 1[b]		Secondary head No. 2
Stub Head	Tertiary No. 1	Tertiary No. 2	
Stub No. 1[c]	Field item No. 1	Field item No. 2	Field item No. 3[d]
Stub No. 2	Field item No. 4	. . .	. . .
. . .	. . .	. . .	. . .

[a,b,c,d]Footnotes in order horizontally.

Characteristically, a table contains no vertical lines. Many computer programs provide an option to use grids in which to plot data. Look at the journals published in your discipline. Unless they regularly publish tables on grids, don't choose this option. Three horizontal lines run the full width of the table, one beneath the caption and any headnotes, one beneath the headings for the stub and the field, and the third below the field and before any footnotes. Other horizontal lines, called straddle rules, run across all the columns of the field items to which the heading above the straddle line refers.

The boxheadings identify items in columns, which should be the dependent variables, and the stubs identify the independent variables for items in horizontal rows. Comparisons between like elements in the data are made down columns, not across rows. Generally, the main boxhead should give meaning to the items in the field. Are those items yields, concentrations, colors, percentages, or other measures? Any subordinate boxhead simply adds precision to the main head. Use as few headings as possible to communicate clearly; only rarely should you go beyond tertiary headings and not beyond secondary heads if possible. Be sure to include units of measure.

The sequence of footnotes is carried horizontally across each entry in the table. For publication, tables are usually typed on pages separate from the text, and a marginal note is made beside the first mention of the table in the text so that it can be positioned appropriately in the layout for the printed manuscript.

Excellent information on tables is presented by Day (1994), the *CBE Style*

Manual (CBE, 1983), and Smith *et al.* (1980). See also the example in Appendix 8. But for the specific stylistic details for your discipline, consult your journals and style manuals. Unless your style sheet specifies otherwise, you can use the following guidelines.

PREPARING TABLES

For Publication

1. First study the publisher's style sheet carefully and look at examples in the text.
2. Type each table on a separate page with double spacing throughout.
3. Use arabic numerals to number the tables.
4. Prepare tables with the like items, or items you want to compare, reading down columns, not across lines.
5. Group items logically with control values to establish a base line for comparisons. Any gradation or trend in data should probably be stressed.
6. Round off numbers. Don't use excessive decimal places. Decimals must be aligned in columns.
7. Be sure the caption is descriptive of the table's contents. No verb is necessary in this caption.
8. Explain in the caption, in headnotes, or in footnotes all nonstandard abbreviations and symbols used.
9. Verify all information. Make sure that all data and statistical analyses are accurate. Verify again after any revision or transfer of information.
10. Check tables for accuracy in use of symbols, units of measure, and other labeling. Be consistent with such labeling among all tables and figures in the same text.
11. Proofread carefully.

For Slides and Posters

Most of these suggestions apply equally to tables that are used in written reports or for oral presentations. The viewer of a poster or slide presentation is not going to be able to study the table as long as the reader of a manuscript can, and so simplicity is even more essential. In publication, the use of symbols, shapes, sizes, and especially color may be limited, but the slide or poster can make important use of such symbolic language. If you are selective in the media you use and if you present representative data, your tables will lead the reader or viewer to the conclusion reflected by the full data. That same remark could be made about figures.

FIGURES

Illustrations, or figures, come in various sizes and shapes. Photographs, drawings, flowcharts, line graphs, bar graphs, pie charts, maps, and variations of these forms can add an understanding that is difficult to convey in words. In considering where and how to use an illustration, keep in mind the values and standards we are striving to achieve in scientific writing. Consider the purpose and what form best fits that purpose. Look at the nature of the data—numbers, ranges, differences, and appearances—and what point you should make with the data. Try to determine how your audience will respond to the illustration. Will they see something different from what you see because they are not familiar with the information? Be familiar with the conventions used with illustrations including symbolic meanings that are expressed with lines, sizes, bars, numbers, colors, shading, and axes, as well as words. And understand any constraints in producing clear communication, whether that be the reproduction of a photograph, the capabilities of you or a graphic artist, or the limitations of the equipment your publisher or you use. With all these things in mind and with an honest, sincere effort you can produce an illustration that carries a simple, accurate message.

If illustrations are not clear, accurate, and appropriate, they are mere distractions and should be avoided. If you need to display specific values, a table is usually the better choice. A great deal that has been written about charts has to do with problems involved in their use, whether it be the difficulty in presenting the data, the difficulty in interpreting the data, or what Tufte (1983) calls "chartjunk" or others might call "noise." Anything that distracts, anything that is simply decorative, anything that is illegible, and anything without a positive communication value should not be used with illustrations any more than such should be used with words or the data themselves. Be sure that the end result of your presenting an illustration is that the audience concentrates on what is communicated, and not how the picture looks.

Computer technology has made possible quick and accurate creation of figures without the graphic artist. Find a graphics software program that works well for your kind of data, and then use it carefully. Most programs will allow you to draw boxes, shadows, and 3-dimensional bars. They provide some atrocious designs to fill the bars or place them in multiple rows with depth perception. Keep your graphs simple; data are often difficult enough to comprehend without challenging the reader to see bars hidden behind bars or to be distracted or confused by strange and indistinct designs. *Data are best presented in the simplest possible form.*

Designing illustrations and the graphic presentation of statistical data are subjects for entire books, and all of us could well benefit from a course on those subjects. The most that I can do is to introduce you to standards and values that

I have encountered in working with scientific communications and to suggest other sources of information. First, by all means, check with *Illustrating Science* (CBE, 1988). For presentation of simple graphs, MacGregor (1979) is still good, easy to read, and well illustrated. Tufte (1983) has produced an interesting study of visual display of data. His includes some history of the art and discussion of problems such as the inadvertent "lie factor" and chartjunk that cloud graphic communication. He illustrates his points well with examples from published charts. Schmid (1983) also presents concerns about the problems in graphic accuracy and in facility of communication. He uses good illustrations and covers the material on graphs and maps well. Chambers *et al.* (1983) get into the more complex area of the visual display of statistical information. Their text requires some knowledge of statistics. In the biological and medical sciences, illustrations, drawings, and photographs of life forms are often depicted. Hodges (1989) provides good material on such illustrations.

Explore these sources, and you will come away with a basic understanding and respect for standards in visual display. After you have looked into at least two or three of these, look more closely at the illustrations in the journals in your area. You will find errors there, but you will also find what works best for the presentation of your data. Along with the information from *Illustrating Science* (CBE, 1988), consider the following basic principles for producing graphs and other figures. Notice the example in Appendix 8.

PREPARING GRAPHS AND OTHER FIGURES

1. Be sure the graph carries your point better than the text or a table would. If the figure becomes complex and requires extensive explanation, reconsider, divide the data, or try something different.

2. Consider what kind of figure you need. Do you need a photo, a line drawing, or a graph. What kind of graph would be best—line, bar, pie?

3. Make it simple. A figure should be comprehended at a glance. Draw graphs to agree exactly with experimental data, but don't overload them with information.

4. Limit the number of curves or bars on a graph. A single figure cannot satisfactorily accommodate more than 3 to 5 lines or 6 to 8 bars (9 or 10 bars if they are grouped).

5. Plot any independent variable on the horizontal (*x*) axis, or abscissa, and the dependent variable on the vertical (*y*) axis, or ordinate.

6. Avoid wasted space. Scale details to agree with the data, but do not extend the axes beyond the point needed. Put the legend into the field of the graph if possible.

7. Label all axes carefully and show units of measure. Use tics and subtics to subdivide the axis so that you don't overcrowd it with numbers.

8. Most scales should start with zero. If you must compress scales with slash marks or start beyond zero, be sure this modification is clear.

9. Remember that position, size, shape, length, symbols, angle, and color are all visual codes that carry messages to a reader. Do not let them convey the wrong message.

10. If a set of graphs is used in the same paper, poster, or slide set, be consistent and uniform in your use of all visual and verbal codes.

11. Select the size and format to fit the journal or other use for which the visual is intended. (Avoid color for most publications; use color in posters and slides.)

12. For a journal publication, type the caption on a separate page so that the figure can be photographed and the type can be set separately.

Bar Charts

1. Bar charts often have just one measurable axis. If they have two, they are sometimes called histograms.
2. Bar charts can be presented for data collected at even or uneven intervals.
3. The bars should be wider than the spaces between them.
4. Use conservative patterns (solids, shading, or hatch marks) to differentiate bars. For posters and slides, use color.
5. Show significant differences with a least-significant-difference bar or with letters or asterisks above bars.

Line Graphs

1. Line graphs should have two axes. Avoid a third axis if at all possible.
2. Simple line graphs should present data collected at regular intervals to show trends without extrapolation between data points.
3. The curves themselves should be the boldest lines. Axes and tic marks should be less bold.
4. Be careful with line patterns. Dots, hyphens, or slashes can become confusing. Sometimes it is better to use distinct symbols (●, ■, ▲) rather than line patterns. Use distinctive colors for lines on posters or in slides.
5. Plot the length of intervals on both axes so that slopes are not excessively flat or steep.

Although bar charts and line graphs are probably the most common forms for figures in scientific journals, other illustrations are also useful. Maps, flowcharts, diagrams, line drawings, chromatographs, photographs, and other forms can sometimes communicate much more than the text alone. *Illustrating Science* (CBE, 1988) is a good source of information on all these illustrations, and Imhof (1982) is a master of cartography. Hodges (1989) presents very good information on techniques for creating illustrations as well as the subject matter, especially in the biological sciences. Whatever illustration you use, be sure to submit high quality for publication and for use in slides and on posters. An out-of-focus photograph or a graph filled with chartjunk (Tufte, 1983) will merely distract from what you wish to say. You may also need assistance from a graphic artist for line drawings or other figures, but be sure you understand enough about the illustration yourself that you will know what it is communicating. Symbolic communication with or without accompanying words is a strong form of communication. Be certain that it carries the right message.

References

Chambers, J. M., Cleveland, W. S., Kleiner, B., and Tukey, P. A. (1983). "Graphical Methods for Data Analysis." Wadsworth International Group, Belmont, CA, and Duxbury Press, Boston.

Cleveland, W. S., and McGill, R. (1985). Graphical perception and graphical methods for analyzing scientific data. *Science* **229,** 828–833.

Council of Biology Editors (CBE) (1983). "CBE Style Manual," 5th ed. CBE, Bethesda, MD.

Council of Biology Editors (CBE) (1988). "Illustrating Science: Standards for Publication." CBE, Bethesda, MD.

Day, R. A. (1994). "How to Write & Publish a Scientific Paper," 4th ed. Oryx Press, Phoenix, AZ.

Gastel, B. (1983). "Presenting Science to the Public." ISI Press, Philadelphia, PA.

Hodges, E. R. S. (1989). "The Guild Handbook of Scientific Illustration." Van Nostrand–Reinhold, New York.

Imhof, E. (1982). "Cartographic Relief Presentations." de Gruyter, New York.

MacGregor, A. J. (1979). "Graphics Simplified." Univ. of Toronto Press, Toronto.

Schmid, C. F. (1983). "Statistical Graphics: Design Principles and Practices." Wiley, New York.

Smith, R. C., Reid, W. M., and Luchsinger, A. E. (1980). "Smith's Guide to the Literature of the Life Sciences," 9th ed. Burgess, Minneapolis, MN.

Tufte, E. R. (1983). "The Visual Display of Quantitative Information." Graphics Press, Cheshire, CT.

12

ETHICAL AND LEGAL ISSUES

"Whatever the rationalization is, in the last analysis one can no more be a little bit dishonest than one can be a little bit pregnant."

C. IAN JACKSON

Ethics are determined by cultural, social, and professional values and may not be governed by laws. Because human values dictate ethics, standards and beliefs can differ between individuals, groups, and cultures. Laws supposedly apply the same for all those ruled by the same governments. Yet both laws and ethics result in behaviors important to human interactions. In scientific communication, professional ethics and copyrights and patents are the main issues of ethical and legal concern.

ETHICS IN SCIENTIFIC COMMUNICATIONS

Most scientists maintain high ethical and professional standards. As scientists, they are interested in discovering truths, and any communications among them must be carried out with a high degree of accuracy and integrity. It is in their best interest to protect the pool of scientific knowledge from any fabricated data or conclusions based on insufficient evidence. But human errors do occur, and good things can happen to bad people. Isolated instances of scientific fraud have been found even among previously very reputable researchers. Every scientist is responsible for protecting the integrity of science.

In scientific communication, two kinds of ethical errors are unforgivable—distorting your own data and plagiarizing the work of others. Other breaches of integrity are also deplorable but sometimes hard to diagnose. Most difficult is recognizing your own bias and admitting that even you can rationalize truths. It is not always easy to mark the dividing line between truth and assumption, and all of us can err. Such errors can be made inadvertently with no intent to lie, but for that reason they are all the more important and all the less forgivable. You have to discipline yourself to demand care, accuracy, and objectivity in every instance.

It's easy to see the dishonesty when you change numbers in a data column, but what of the situation in which you believe a result should occur, the data almost point in that direction, and to delete one experiment would take care of the *almost*. Can you delete that experiment? Was that odd data set simply an anomaly that should not be considered anyway? Is there a typographical error or erroneous transcription in your data? Do you have time to experiment further to substantiate what you believe should have resulted? These questions are not easy to answer, and often no one except you can answer them. You can become so overly meticulous that you are ineffective, or you can rationalize your way into really dishonest reporting. Your scientific integrity is based on careful scrutiny of your work and sound judgment in how you answer the questions involved.

The Committee on the Conduct of Science (1989) concludes that "people have a tendency to see what they expect to see and fail to notice what they believe should not be there." Individual and cultural prejudices have troubled scientific research and reporting for centuries, and we are not without prejudices today. In *The Mismeasure of Man,* Gould (1981) exposes the social prejudice that has produced "documented" studies on biological determinism to prove that some groups of people are inferior to others (women to men, for example). As "proof" these studies have used selected criteria that fail to reflect a total physical reality. Gould's (1981) conclusions should bring a degree of humility to all of us: (1) "no set of factors has any claim to exclusive concordance with the real world," and (2) "any single set of factors can be interpreted in a variety of ways." Think about those remarks when you conclude that you have discovered "truth."

Unintentional dishonesty can also come in communication. We know when we speak or write an outright lie, but we must also be sure our statements are not misleading and ambiguous. A major difference between scientific writing and creative writing is that in creative writing we can allow the readers to interpret as they choose. Ambiguities and double meanings can simply add interest. We are not allowed that avenue in scientific communication. Read your sentences carefully to be sure that they will not be easily misinterpreted. The same is true for tabular or graphic portrayal of data. An inaccurate number or a visual deception in a graph can misdirect a reader. Your responsibility is to know the conventions for data presentation and check the accuracy of all details. Careful research, use of scientific reasoning, an open mind, clear and accurate communication, and a willingness to be honest at all costs will generally result in ethical conduct. Settle for nothing less in yourself and your colleagues.

Plagiarism is a dirty word in scientific writing, and it is both a legal and an ethical issue. A dictionary will define the term as some kind of literary theft or stealing. Like other forms of dishonesty, plagiarism is easy to distinguish when it is blatant. You lift words from someone else's work without giving due credit. That's clearly plagiarism. But how about ideas or words that are not quite the same but mean the same thing? Or how about picking up from your subconscious something you have read, thinking it is your own idea? These nebulous situations occur, and we cannot always be afraid that we may be using ideas that are not entirely original, but we must give due credit for both words and ideas. To guard against plagiarism or a failure to document sources, be familiar with the literature in your discipline.

Dartmouth College's Committee on Sources (1988) has outlined three types of plagiarism. The Committee lists not only "direct quotation or word for word transcription," but also "mosaic or mixing paraphrase and unacknowledged quotation" and "paraphrase and/or use of ideas." These three forms are all bad, but it may be that some students and professional writers do not recognize the last one as plagiarism. Of course, you can paraphrase limited amounts of someone else's work if you give due credit, but just putting someone's ideas into your own words without citation and documentation will constitute plagiarism as surely as if you had used their words.

Plagiarism can also result from simple ignorance in how to reference other works or in sloppy, careless documentation. Careful documentation and verification of your references as well as clear knowledge of what has been done in your area of study are essential. Learn the accepted conventions for documenting the work of others and for obtaining copyright permission. When you question your right to use the words or ideas of others, find answers to those questions before you proceed. Some answers may be found in this text, but when questions remain, go to other sources. Ignorance is no excuse.

PROFESSIONAL RESPECT FOR OTHERS

Other questions of professional ethics have to do with your relationship with other scientists and with any professional group to which you belong. Candid sharing of scientific findings and maintaining the confidence of others are essential to your being an ethical scientist. The society or registry in your discipline probably has a code of ethics that you can study and use as a model. Be conscious of what constitutes fraud, conflicts of interest, and the need for confidentiality. Your colleagues soon learn to what extent you can be trusted.

Giving due credit to others is important in your reference to their work and in recognizing colleagues as coauthors. It is equally important that undue credit not be given. Questions of junior or senior authorship and of who should be included as an author are not always easy to answer. If you are a graduate student or a junior scientist, discuss with your advisor authorship and the order of authors on a paper or presentation. The *CBE Style Manual* (CBE, 1983) says that "The basic requirement for authorship is that an author should be able to take public responsibility for the content of the paper." That means that any coauthor should be clearly involved in the research and composition to be able to answer questions asked about methods, results, or any other idea expressed in the paper. Despite what some may call honorary collaboration or honorary authorship, any author listed on a paper should have contributed substantially both to the work and to development of the manuscript or presentation. Giving false authorship a nice name doesn't make it any more ethical. For information on authorships and making ethical judgments, read Macrina (1995) or *On Being a Scientist* (Committee on Conduct of Science, 1989).

Ethics is a matter of making good decisions about questions of appropriate conduct regarding your work and that of others. There is often no legal retribution for offenders of codes of ethics. But, as Day (1994) says, the ethical is more important than the legal. Your colleagues and your professional society may provide some guidelines, but unfortunately no definitive rules regarding ethical judgments exist. Almost every situation demands an individual consideration, and you must draw your own conclusions. Consider the following issues that you will probably encounter in your career as a scientist.

Respect Your Data

Don't make up your mind about your results before your data are collected and analyzed (Fig. 12-1). Trimming, cooking, or otherwise manipulating data is as bad as reporting false data. Be careful with your own prejudices, and be cautious when you work with others, especially with overly specialized scientists. In working across disciplines, researchers have to trust each other for ideas and analyses that their own disciplines do not include. The same is true with specialized statisticians who often know little about your science but can plot and analyze your data in a variety of ways. They can contribute invaluable expertise to a project, but they may not recognize the relative importance of your scientific variables. You need to understand links between your work and

FIGURE 12-1
Respect all of the data you collect, not just select ingredients.

that of your colleagues and to be familiar with the meaning of any statistical analysis used with your data.

Be Careful with Confidentiality

In your own laboratory, information may be available that should not be disclosed to others until experimentation is complete and researchers are ready to publish their findings. Or you may review a paper for someone and find interesting information that should be held in confidence until that paper is accepted and published. At professional meetings, scientists should certainly exchange ideas, but take care in making positive pronouncements on unsubstantiated results. Day (1994) is right in saying that a scientific experiment "is not completed until the results are published." Disclosure before that time should be handled with caution.

Don't Publish the Same Thing Twice

To publish the same data in more than one scientific outlet is unwise. If you need to convey the same information to two audiences, e.g., the scientific community and the public, two publications might be based on the same experimentation but would be presented in different ways. To publish results of the same research in two scientific journals is unethical even if you completely rewrite the text. Too much scientific literature already exists to ask an audience to read the same thing twice. Most scientific journals require that you not even submit the same paper to another journal while it is under consideration for publication. Be sure one publisher has released your manuscript before you ask another to consider it.

Acknowledge Your Errors

You are human; you can make mistakes. If, in all sincerity, you say or write something that you find later is misleading or untrue, try to right the situation by acknowledging that error. To make a mistake is one thing; to allow people to continue to be deceived is another. Your colleagues are also human; they make mistakes. And they will respect you far more if you acknowledge an error than if you let it go, even if the intent to deceive was never present.

Support an Ethical Workplace

When you agree to work for a company, an agency, or an institution, you are also agreeing to abide by the policies of that employer. Many of them have specific requirements relative to communication of information. You may need to have approval for any paper published, speech made, or meetings attended. The employer may also trust you with confidential information that you should protect from unwise disclosure. Sometimes your own values may not concur with the policies of your employer. Ethical decisions on what to do can be difficult. Know your employer's policies and abide by them insofar as they are within the law and conform to your ethical and professional standards. If you are uncomfortable with your employer's ethics, you would do well to find another job.

Respect the Time of Others

If you are asked to contribute to a paper as a coauthor, get your part done expeditiously so that you do not delay the publication for other authors. The same is true when you review a paper for others. Get the review completed in a timely manner so that the author or editor is not unduly delayed. You should also be considerate of a fellow scientist's personal and professional time. Lengthy consultation should be a matter of mutual agreement and benefit. If someone generously provides time for consulting about your research, be sure that person is not coerced and that you acknowledge contributions made.

Watch Out for Conflicts of Interest

Like plagiarism, blatant conflicts of interest are easily recognized. Trying to mix a personal relationship with a professional one or overlapping two professional activities can result in conflicts of interest. Any personal or business relationship should not interfere with your work as a scientist and a professional. Business partnerships with an advisor who has some authority over your degree program or research are questionable as is having kinfolks serve on your graduate committee. Remember that there are no definitive rules. A father and daughter could become a highly ethical research team. But treat any possible conflict of interest with great caution. The scientist must avoid any personal bias in his or her judgments.

Be Fair with Your Time and Effort

Conflicts of interest involve personal and professional relationships with others, or you may find issues in your individual work competing for your time, energy, and effort. Scientists can become involved in many activities—research, teaching, grant seeking, professional obligations to society, speaking engagements, seminars, workshops, travel, and myriad other activities. Quality of any one activity can suffer at the expense of the others. To assume a responsibility and do a poor job can be worse than refusing to assume the responsibility. Only you (and perhaps your boss) can determine where your responsibilities lie and how far you can extend your time and energy.

Avoid the Sin of Omission

Refusal to assume a reasonable work load can be equally as irresponsible as trying to do too much. It's hard to believe that you can be accused of unethical actions if you do nothing, but breaches of conduct come with omission as well as commission. As a scientist, you will often be asked to contribute your expertise to a collaborative project. You may be asked to serve on professional committees or to be a reviewer or associate editor for a journal. You will have to be discreet in determining how much you can do and still maintain quality work. You can't yield to every request for your time and energy. On the other hand, to reject all requests to serve your science or your profession is unethical.

Watch the Company You Keep

You may be Honest Abraham, but if you collaborate in your research with unscrupulous individuals, their reputations will rub off on yours. Be alert to how your colleagues are received in the scientific community. And as you work with a fellow scientist, observe the care he or she takes with accuracy and the truth. You can usually recognize unethical behavior if you are conscientious.

If you have not had an opportunity to take a course in ethics, consult a text such as those of Bayles (1989) and Macrina (1995) and be sure you can define ethics relative to your values and those of your profession. Professional reputations and scientific values depend on scientific integrity. Objectivity and scientific methods are only as honest and unbiased as the researcher or communicator who uses them. Beyond a reasonable point in the critical review processes, we have to trust each other to be honest and accurate. Generally speaking, scientists are honest people. Those who cannot be trusted in their experimentation or reporting have no place in the world of science. But devils do exist, and we must be constantly alert to unethical and unprofessional behavior. The least you must do is to control your own behavior and uphold the highest ethical standards. As Thomas Carlyle said, "Make yourself an honest man, and then you may be sure that there is one rascal less in the world."

THE LEGAL ISSUES: COPYRIGHTS AND PATENTS

Sometimes the message you want to communicate to other scientists is a simple "It's mine." Or you may wish to make use of a creation that belongs to someone else. In either situation, you are dealing with legal, as well as ethical, issues. Laws governing copyrights, patents, and trademarks protect the property of individuals and groups. Scientists must use each other's ideas and inventions, or we wouldn't make much progress, but be sure you give credit to your sources and ask permission when you make extensive use of the work of others. You will also be asked to grant permission to others. Know your legal rights, but recognize the need to exchange information with other scientists. A few notes on copyrights and patents may be helpful.

Copyright

You own the copyright on any tangible expression that you create. Along with other forms, tangible media include written words, illustrations, printed works, electronic software, and recordings. Copyright does not cover the ideas, procedures, processes, concepts, or discoveries contained in such works. Protection of those things can be obtained through patents. Copyright registration is not a condition required for copyright ownership. Statutory copyright begins when the document is created, but protection and use of your work are increased when you have the copyright registered. If a work is not yours, find out whose it is before you use it.

The author involved with scientific communications needs to be conscious of copyright privileges and regulations. You will need to know how to register copyright for yourself, how to grant permission, and how to obtain permission for use of works from other copyright holders. Below are some handy things to know, but consult the copyright law or a lawyer when you have important questions on copyright.

Both published and unpublished works in any tangible medium of expression are protected by copyright and may be registered with the Office of Copyright. Whether it is registered or not, copyright protects only the expression of an idea, not the idea itself.

Registration of copyright can be obtained any time during copyright duration. For information on registration of copyright, you may write to the Copyright Office, Library of Congress, Washington, DC 20559. For written scientific work, you will want the forms for registration under Nondramatic literary works.

Duration: In general, works produced before January 1, 1978, were already under protection and usually can be protected for about 75 additional years from that time. On or after January 1, 1978, works are under copyright until the death of the author plus 50 years thereafter.

Works for hire are those expressions or inventions that an employer requires or commissions you to create as a part of your job. In other words, you are being paid to create something, and any copyright, patent, or other benefit that comes from the creation belongs to the employer.

All works prepared by officers or employees of the U.S. government as part of their official duties are in the public domain. You don't have to obtain permission to use these materials, but you will still document your source.

"Fair Use"—not clearly defined—generally means use for your own educational purposes as opposed to use for commercial gain. This definition is certainly oversimplified and incomplete. If you question your rights to use the works of others, consult an authority or apply to the copyright holder for permission to use the work.

Electronic communication has added a new dimension to copyright. Software bought by one user should not be copied by another any more than a book bought by one should be copied for a friend or colleague. New laws are being considered for how to copyright and document electronic communication. Be alert to developments in this area.

To Grant Copyright Permission From journals that publish your work, you may receive a form that you and any coauthors will sign to transfer copyright or to grant printing and reprinting rights to the journal. Some publishers may ask that you transfer the copyright itself so that they are the owners, and you must refer any request for permission to use the work to them. Others are simply asking that you grant them permission for reprinting and distributing the work. In this case, you still own the copyright.

If individuals ask for permission to use your work or a part of your work, they may send a form to be signed or a letter requesting permission. Be sure exact portions of works to be used are fully described or copied and that the description or copy is kept with the signed permission granted. Keep this information on file along with specific details on where and when the material will be used. Request that persons using your material clearly acknowledge both the source from which it came as well as you and any coauthors.

To Obtain Copyright Permission When you know who the owner of the copyright is, write directly to that person or agency for permission. You should begin efforts to obtain any copyright permission as soon as you know that you want to use the material. You may want to make a phone call to locate the copyright holder and to see whether he or she is amenable to your using the material. However, a phone call will not constitute evidence that you have been granted the permission. You need specific information in writing, and this process may take weeks or months. You can delay your own publication by waiting too late to apply for copyright permission.

In requesting permission to use another author's or publisher's work, clearly identify the specific material to be used, including

1. Author(s)
2. Title
3. Date of publication
4. Publisher (if your letter is not directed to this agency)
5. Specific selection to be used
   A. The form (as table, photograph, or text)
   B. A description of the content (perhaps include a photocopy of the material you will use)
   C. Pages on which it appears in the original text

Tell the agency or person to whom you apply where, when, and how you will use the material requested; whether your work will be published and what form it will take; and who the audience will be. Indicate that you will give full credit to the author(s) and publisher, and be sure that you do so. You will often use a form letter in obtaining copyright permission and will want the information in duplicate for your files as well as the grantor's. Ask for signatures of all authors. Keep a clear record of permission received.

A sample form letter or other information on copyrights is found in the *CBE Style Manual* (CBE, 1983) and in Dodd (1986). Appendix 9 also provides an example for requesting copyright permission.

Patents

Patents are more versatile and offer more protection than copyrights where they are applicable, but they are not applicable to the expression of ideas.

Patents protect the ideas as they are put into practice as machines, manufacture, processes, or composition. Copyright protection begins as soon as the expression is created; patents must be registered to serve as protection. Copyrights protect for longer periods than the 17-year protection for patents. Work for hire is applicable to both copyrights and patents. The remarks here on patents apply just to U.S. patents.

For the sake of simplicity, I refer here to any patentable creation as an "invention," but patents can be obtained on diverse items including synthesized materials and life forms. As with copyright for electronic communications, biotechnology as added a new dimension to patents. It is now possible to patent life forms if they have been synthesized by human efforts and do not exist in nature without the human intervention. If you are interested in patents governing life forms, read Macrina (1995), especially Chapter 9 by Cindy L. Munro.

Any invention that is patented must be proven to be novel, not obvious, and useful. In applying for the patent, you must give evidence that your invention has all three of these characteristics. The three overlap in meaning. **Novel** simply means that your invention is new, that a like invention did not exist before yours. **Nonobvious** means that what you have put together is not something that anyone would immediately derive from the same materials. Your invention is needed, but no one else has come up with a way to fill that need. Finally, in filling this need, you have made something that is **useful.**

Several communication efforts are important relative to patents. First, you must disclose to the Patent Office any information you have on how the invention works, its components, and any similar inventions already in existence. These disclosures require that you carefully search and read the literature on your subject. Any failure to disclose similar inventions already in existence can thwart your chances of obtaining a patent even if you did not know about the existing invention.

An important question relative to communications and patents is that of when to publish information about your invention. If you describe your invention to the public before you have registered the patent, it is considered public information and will probably not be patented. Be careful about presenting a paper on your invention at a professional meeting, publishing a journal article about it, or even describing it to anyone beyond confidential disclosure to colleagues who can be trusted to respect your claim to ownership. Although the same is not true in many other countries, in the United States if you do publish information about your invention before you realize that you should obtain a patent on it, you can apply for a 1-year "grace period" for filing the patent application, but remember that you must be able to prove that you invented it first. Someone else may already be using the idea or even applying for patent. Records that you keep as you are working on the invention can be essential to proving that the invention is yours. Keep careful, dated records, and if you foresee that you might be seeking a patent, have witnesses sign and date information that you record along the way.

Finally, perhaps most important to scientific communications is the patent literature. The disclosures on how inventions are produced and how they function are published and offer a storehouse of information helpful to other scientists. Paul (1975) claims that "The U.S. patent literature is the largest and most comprehensive collection of technical information in the world." Certainly, the information on patents contains useful ideas, and you should not neglect to explore your research subject in this area.

If you are thinking of patenting an invention, you are sure to have numerous questions that I cannot answer. Complexities arise in the patenting process that often require the services of a lawyer specialized in the area. In fact, one of the first things you may want to consider is hiring a patent attorney to deal with the Patent Office in proving that your invention is novel, nonobvious, and useful. In addition to that of Macrina (1995), other books on the subject may also prove useful. Grubb (1986) has good information on patents in chemistry and biotechnology, or Crespi (1982, 1988) can introduce you to patent processes in the United States and in other countries.

References

Bayles, M. D. (1989). "Professional Ethics," 2nd ed. Wadsworth, Belmont, CA.

Committee on the Conduct of Science (1989). "On Being a Scientist." National Academy of Sciences, National Academy Press, Washington, DC.

Committee on Sources (1988). "Sources: Their Use and Acknowledgment." Dartmouth College, Hanover, NH.

Council of Biology Editors (CBE) (1983). "CBE Style Manual," 5th ed. CBE, Bethesda, MD.

Crespi, R. S. (1982). "Patenting in the Biological Sciences." Wiley, New York.

Crespi, R. S. (1988). "Patents: A Basic Guide to Patenting in Biotechnology." Cambridge Univ. Press, Cambridge, UK.

Day, R. A. (1994). "How to Write & Publish a Scientific Paper," 4th ed. Oryx Press, Phoenix, AZ.

Dodd, J. S. (ed.) (1986). "The ACS Style Guide: A Manual for Authors and Editors." American Chemical Society, Washington, DC.

Gould, S. J. (1981). "The Mismeasure of Man." Norton, New York.

Grubb, P. W. (1986). "Patents in Chemistry and Biotechnology." Oxford Univ. Press, New York.

Macrina, F. L. (Ed.) (1995). "Scientific Integrity: An Introductory Text with Cases." ASM Press, Washington, DC.

Paul, J. K. (1975). "Fruit and Vegetable Juice Processing." Noyes Data Corp., Park Ridge, NJ.

13

SCIENTIFIC PRESENTATIONS

"Nothing clarifies ideas in one's mind so much as explaining them to other people."

VERNON BOOTH

Spoken presentations at professional meetings are extremely important to scientific communication and to your individual reputation. You will probably be required to make one or more presentations while you are in graduate school. You may even have a course in your department in which you and other graduate students regularly deliver speeches or slide presentations. Also, take advantage of opportunities to attend seminars and special lectures in other departments. Valuable experience can be gained by attending presentations either as a speaker or as a listener.

If you have occasion to attend regional or national meetings sponsored by societies allied with your discipline, go and present posters or slide talks. In doing so, you are establishing your reputation with colleagues, and you may encounter prospective employers for whom you will later make a presentation at a job interview. Whether it is a departmental seminar, participation in a national meeting, an informal or formal speech, or a job interview, you must perform well when you are "on stage" to become recognized as knowledgeable and articulate, two qualities almost essential to your success as a scientist.

DEPARTMENTAL SEMINARS

A specific time, perhaps weekly or semimonthly, may be set up in your department for seminar presentations. Graduate students and other speakers will be scheduled to present information on their research, on the scientific literature, or on other subjects of interest to the department. During your presentation, keep in mind that you are talking to the people who know you best, but they are also the people who will or will not recommend you for career positions. Whether it is evident or not, professors evaluate your research and your communication skills when you present a departmental seminar. Attendance at these seminars is usually considered one of your professional or academic responsibilities, and participation offers you several advantages.

Seminars Provide Information on Current Research

For both the speaker and the audience, seminars present a unique educational opportunity. Every discipline includes a broad range of subject matter. You cannot expect to become proficient in all areas, but you can obtain some knowledge of and respect for the work being done in specialized areas other than your own. As you prepare a presentation, your perception of your own specialization will increase. Whether you are listening or speaking, seminars furnish a painless way to broaden your education.

Seminars Provide New Perspectives for Your Own Work

From others who discuss their research proposals, methods, and results, you will acquire ideas that apply to your own work. Seminars present a means for uncovering errors, picking up new perspectives, and strengthening your own research. Constructive, professional criticism is always beneficial for both the beginning scientist and the experienced professional.

Seminars Increase Your Ability to Evaluate Research

As a graduate student, you soon learn that no scientist is infallible. Much time, effort, and objective criticism are required to judge whether a scientific paper or presentation reflects good research, whether it is presented well, and whether it contains significant new ideas. The ability to listen to and critically evaluate a presentation is useful in acquiring new ideas and in deciding what information is valuable for your own research or communications.

Seminars Improve Your Ability to Communicate

Education and scientific progress are so closely allied with personal communication that everyone involved needs to develop an ability to communicate well. Few can become effective speakers without conscious effort. During graduate school, you will not make enough presentations to provide the desired training, but put forth your best efforts when you give a seminar and critically observe the efforts of your colleagues. The experience you gain will be well worth the effort when you deliver a presentation at a professional meeting or a job interview.

THE PROFESSIONAL MEETING

Communication at scientific meetings transpires mainly through the spoken word used formally and informally. The best information from meetings often comes from casual conversations. More formally, the speaker is a valuable part of the poster or slide presentation. When you are talking about science and research, you need to maintain your professional attitude whether you are in a formal or an informal situation.

The importance of chance encounters and casual conversations or the role of an author at a poster should not be underestimated, and some preparation can be made for these exchanges. Basic to good communication here is that you know your own material and the literature on the subject. Know how your research was planned, designed, and carried out; how data were collected and analyzed; and how your results compare with those of others who have done similar research. You may even want to take data or notes to the meeting to verify remarks you make. Otherwise, prepare by going over your material before you get to the meeting and try to predict what questions might be asked about it. To get the most from attending a professional meeting, you should be prepared for your own communication and plan ahead for the activities you will be involved with.

Getting the Most from a Professional Meeting

—Study the program that is usually published well before the meeting, and plan your own schedule.
—Plan to give a slide or poster presentation. Enter a contest with your work if one is held.
—Carefully select other presentations to attend including those by authors whose publications you have read.
—Unobtrusively critique the good and bad points in posters and slide presentations to apply to your own research and communication.
—Observe the leaders of your society and how they conduct the business of the meeting.
—Take advantage of placement or career services. Meet as many new people as you can.
—Schedule time to relax and enjoy highlights of the town with your friends and new acquaintances.

Presentations at Professional Meetings

Relative to your career, the highlight of the professional meeting is your presentation of a slide talk or a poster. Both of these formats are prominent at meetings, and both require your skills in speaking. You must keep the audience, the subject, and your visual aids in mind as you talk about your work. You will often be required to make the decision on whether to present a poster or a slide talk. Try to get experience with both formats. Consider characteristics and requirements for both and choose on the basis of which best fits your material and your ability. The comparisons and contrasts in Table 13-1 may be helpful.

TABLE 13-1
Comparison of Characteristics and Requirements for Slide and Poster Presentation

Poster presentations	Slide presentations
The Situation	
Relatively informal; contact one to one or one to a few	More formal; contact one to many
Both speaker and audience standing	Speaker standing and audience seated
No moderator; direct contact with no buffer between the speaker and audience	Moderator helps to introduce, buffer the audience, and keep time
Time limit flexible	Time limit formalized
Audience free; only the truly interested remain	Audience more captured; most not likely to leave
Chiefly question and answer or conversational discussion	Chiefly declamation from speaker with a short question session
Handouts helpful; easy to exchange names and addresses	Handouts possible; not likely to exchange names and addresses
Preparation	
Materials: Poster, tacks or Velcro	Materials: Slides, carousel, and notes if needed
Know your subject—Be able to justify objectives, refer to literature, and support your methods and results	Know your subject—Be able to justify objectives, refer to literature, and support your methods and results
Prepare answers to the questions you expect	Prepare formal speech to adhere to time limits
Get ready early. Construct poster, review, and revise	Get ready early. Practice, review, and revise

Remember that most of the people in your audience will know less than you about your subject. Prepare your talk or poster for these people. Try to develop the clearest and most effective way to explain the subject to them. If you aim your presentation at the few people in the audience who know more than you about the subject, you may succeed in convincing them that you understand your material, but likely you will simply obscure the real significance of your subject from most of your audience.

Even with your own research peers, avoid use of statistical and technical jargon, but indicate what statistical analyses have been applied to your data. In most scientific research, statistical techniques are used only to provide a test of significance or to obtain an empirical mathematical expression of relationships. Emphasize the fundamental scientific concepts, not the statistical techniques. If you must use jargon terms peculiar to your subject, define them clearly for your audience.

Try to orient your talk or poster around **one central idea.** Accept the fact that everyone in your audience will forget most of what you say, but if you do your job well, most of the audience will remember you and your point of emphasis for at least a few days. If you fail to distinguish between big points and little ones, your audience will not make that distinction. An audience will simply walk away from a confusing poster and a confused author. The speech audience may sit in their seats as a matter of courtesy, but their minds will have turned to more interesting subjects. Also keep in mind that they are there because they are interested in what you have to say. Present your basic points vividly. Restrict the scope of your subject, and don't leave out a thorough explanation of the essential points.

For clear communication, you must be conscious of symbolic communication and communication without words. Your attitude, facial expressions, tone, and all the symbolic displays in your slides or poster may carry a stronger message than anything you say. Visual aids, properly prepared and used, can enhance most presentations. At their best, however, visual aids are merely aids. At their worst, they can completely destroy the effectiveness of your presentation. They do not substitute for adequate preparation and effective verbal exposition. For the poster they should support and illustrate the written material and your comments or responses to questions from viewers. For slide presentations, consider yourself and your speech content, not your slides, the central focus of the presentation.

SPEAKING AT THE JOB INTERVIEW

You are even more specifically the central focus when you make a speech at a job interview. Taking a slide presentation to a job interview may appear somewhat more frightening than making other presentations, but it need not be. If you have taken every advantage of experience in departmental seminars and at professional meetings, you should have built the confidence that can make you a good speaker. Take that confidence with you to the interview.

For the speech at the job interview, two points deserve special attention: the audience and the purpose of the speech. For your departmental seminar, your **audience** is made up mostly of people you know, and many of them know what kind of research you have been doing. You are usually given enough time to fully explain your points. At the professional meeting, your audience is a group of people especially interested in your topic; they probably already know much about it. Your time is short, but they don't need a great many details to understand a point you want to make. Most of them are more interested in your research than in you. The same is not true with the job interview.

The audience at the job interview is often made up of administrators, managers, and scientists with diverse backgrounds who are interested primarily in finding out more about you. There may be few with any expertise in your area of research. You must present your material so that it is clearly understood by individuals who are probably highly intelligent but who are uninformed about your subject. Because the audience is different from those in your department and those at the professional meeting, don't expect to make the same presentation to them that you used for the other occasions. You need to revise a presentation every time you present it and especially when a job is at stake.

Because the audience's chief interest is you, you must align the **purpose** of your talk with that situation. They want to know whether to hire you, whether you will work well with them, and whether your expertise fits the position they need to fill. They want to know whether they will enjoy a professional association with you. Your purpose then is to provide positive answers to such questions. To do so, simply follow the principles for all good presentations, but alter your own approach to accommodate the different set of questions.

Give your audience the opportunity to discover more about you than is evident in a research presentation. For example, after you are introduced, it is a good idea to leave the lights on for a minute or two, thank the audience for inviting you, and give a brief but not overly zealous explanation of your interest in the job. Then use a transitional remark such as, "My interest in this kind of work has increased with my research on. . . . Today I'd like to show you one part of that research in which. . . ." At that point you are ready to turn off the lights and turn on the projector. This brief interlude between your being introduced and your presentation can put the concentration on you and can add immeasurably to the audience contact.

As with any presentation, maintain good eye contact throughout the talk. If you have a choice, keep the speech itself short, no more than 20 to 30 minutes; explain fully a limited number of points; and relate your study to that of other researchers. Some of them may be in your audience. Establish credibility with your experimental design and analyses, and report results for which you have strong evidence. This is not the time for you to speculate on momentous breakthroughs that you believe you have made in science. That is not to say that you should be overly modest. Show the audience an example of the good work that you have done, and invite questions so that you can provide answers that will establish your expertise. Your attitude should be that of any good speaker—confidence flavored with a good dash of humility.

Limit the amount of material you present. The mistake I see made most often is that scientists seem to believe that they must display all the data they have collected and analyzed over a period of several years. Be selective; present only a part of your study. A few years ago, I watched a former student of mine interview for a position. He was given 30 minutes for his talk, and he unwisely decided to present work that he had done since he had received his doctoral degree, plus a segment on different work he had done for the doctorate, and then still another subject from the data collected during his master's degree. When I asked him why he had presented so many studies, he said that he believed the audience would be more impressed with the amount of work he had done than with the details. I disagree. The audience can assume that he has done a great deal of work in earning two degrees and holding a responsible position for 3 years thereafter. What they can't do is digest three complex studies in 30 minutes. The speech had loose organization to accommodate all three studies, it ran beyond 30 minutes, and there was not time to demonstrate credibility with details and show quality research. (The young man was not offered the position.) Audiences are accustomed to time limitations. Establish your credibility, present quality methods and results, and show the relationship between your work and that of others. The audience can then readily assume that you have done other work of the same caliber.

One last word on presentations at job interviews. Learn all you can about the position, the location, and the people in your audience before you get to the interview. Your major professor or other advisors can be helpful in providing background information and preparing you to go to the job interview. Your knowing what kind of research is being carried out, even if it is unrelated to yours, and your knowledge of the work that particular scientists there are doing can influence whether you are offered a job. Don't be surprised if, after your talk, the listeners seem rather uninterested in your research and ask questions that have no relationship to what you've been talking about. Remember that their concentration is on you, not your research. For them, your talk serves as a critical demonstration of what you can do and of how articulate you are. They already have a resume, transcripts, and letters that reveal your experience and abilities. They are now checking out a personality. Recognize that point and deal with it positively before, during, and after your talk.

THE QUESTION AND ANSWER SESSION

Closely related to any formal presentation are your interactions with an audience during the question and answer session. Don't underestimate the importance of these interactions. The question–answer session allows the audience to clarify points or add to their knowledge of your subject. It can also build your reputation as a scientist and speaker, and it provides you with an opportunity to surmise the strengths and weaknesses in both your research and your delivery by the kinds of questions asked and your ability to answer them. You must keep the entire audience in mind during the question–answer session. Preparation for the session requires that you know your subject and maintain your confidence.

Give clear, concise answers. Don't dismiss any question without a response, but don't belabor any point. Any question is important even if it sounds trivial. Don't allow yourself to be pulled into a controversy. You probably know your subject better than most of the people listening, but the time and place are completely wrong for any heated disagreements. After the presentation, you may want to continue a discussion with some individual, but do so only after you have released the audience.

Most people will not interrupt you during your talk, but if someone does, don't panic. Answer his or her question or respond to a remark courteously and completely but as briefly as you can. Keep your place in your own presentation (probably via your notes or slides) and return to your prepared speech as quickly and smoothly as possible.

Above all, maintain a professional attitude throughout the question–answer session. Many speakers tend to lose their professional demeanor when the last note on their conclusions dies down. They may loosen a tie or lean on a podium and relax their diction too much. "Yeah" is not a good way to begin the answer to a question. Avoid these distractions and maintain your role as speaker. If possible, let the moderator make the transition between your speech and the question–answer session. This technique gives the speaker a chance to relax momentarily. Haakenson (1975) provides some good information on handling the question and answer session. The following suggestions may help you also.

1. Listen closely. You cannot answer well without hearing and understanding the question. Don't interrupt before the question is completed even when you know what is being asked.

2. Repeat the question aloud if there is even a remote chance that it was not heard or is not clear to you or the audience.

3. Pause. There's nothing wrong with taking 2 or 3 seconds to think, and your answer will probably be better for it.

4. Answer the question completely but as briefly and directly as possible. Don't go into a new speech. Others may also have questions. If

beyond a reasonable answer, suggest to him or her that you meet to discuss the matter further after the session.

5. Don't be afraid to say you don't know. Questions may be asked that are only remotely related to your subject. Simply indicate that your research has not supplied an answer to the question. Refer to the literature if you know a source for an answer, but don't guess.

6. Reply courteously to all. Accept statements and "loaded" or trivial questions and supply a professional comment in your answer. You can often dignify a question that was not presented with dignity by supplying a serious, professional answer that is, at least, related to the subject.

7. Never lose your dignity. Anger is the easiest way to lose it. The audience will have increased respect for you if you reply to the hostile question with a smile and a serious answer.

8. Don't speak beyond your time limit. End the questions if the moderator does not do so, and make a final summarizing statement if possible.

ROLE OF THE MODERATOR

Whether the occasion is a departmental seminar, a speech at a professional meeting, a job interview, or some other speaking situation, the speaker may need to coordinate efforts with other speakers, a program coordinator, a slide projectionist, and especially a host or moderator. The speaker should arrive at a meeting early and meet the moderator and perhaps the slide projectionist. If more than one speaker is on the program, the projectionist needs to know who you are and when your speech is scheduled so that he or she can have your slides ready. Let the moderator have any information he or she needs to introduce you, and be sure to coordinate your efforts relative to lights, time signals, and the request for questions.

On the other hand, you may some day be the moderator and chair an entire session at a professional meeting. Be sure that you can pronounce the names of presenters and the words in their titles. Your job is to introduce them, help them feel comfortable, and solve or buffer problems that arise. In chairing a session, you should provide transitions from one presentation to the next, and be sure that you keep everyone on schedule so that one speaker does not encroach upon the time of another.

To Be a Moderator

1. Obtain copies of abstracts of the talks you are moderating and familiarize yourself with each topic. Prepare a few relevant questions for the speaker to get the discussion started if the audience does not.

2. Help the speaker in arranging visual aids or needed equipment. The moderator should coordinate the operation of lights, projectors, and other equipment with the speaker and should be present in plenty of time before the session to assist with last-minute details.

3. Talk with the presenter or obtain a brief resume and prepare a short introduction of the speaker and his or her topic. This introduction will probably include

a. Speaker's name and title
b. Academic and professional background
c. Any special distinction
d. Title of presentation

At professional meetings, don't use up the speaker's time. You may introduce him or her simply by name, institutional affiliation, and the title of the presentation.

4. Keep up with time. For example, with 20 minutes provided we might expect the speaker to talk for 16 (plus or minus 1 minute) and answer questions for 4 minutes. The moderator must be responsible for keeping everyone on schedule; i.e., see that speakers start on time and finish on time.

5. Be sensitive to problems the speaker may have. Check equipment and know where replacement elements like bulbs are located. Coordinate all efforts with the speaker. Buffer him or her from a hostile question or a string of questions that do not allow the speaker to move on.

6. Accomplish all your responsibilities in a congenial and professional manner.

FIT THE OCCASION

Scientific presentations can take numerous forms other than those described here for departmental seminars, professional meetings, and job interviews. In these situations as well as presentations to such groups as public school children or civic clubs, you may make a speech without visual aids, provide a demonstration of a scientific reaction, host a video or film presentation, or serve as moderator for a symposium or a group discussion. Equipment may dictate that you use overhead transparencies rather than slides. In the future, equipment at many meetings will likely allow you to project slides and videos onto a screen by signaling through a computer via a keyboard or mouse. Adapt to the situation, but keep basic principles of clear communication in mind as you make use of new situations and alternative media.

In the following chapters you will find more specific information on visual, verbal, and symbolic communication used in speech making, slide and poster presentations, and group communications. Most of the decisions on how best to communicate rest with you, but knowing the expected conventions can serve you well in making these decisions. Communication is both a personal and a social activity. Be creative, but rely also on standards or conventions that everyone uses. In other words, be yourself and use techniques that serve your personality best, but satisfy the expectations of your audience by using conventions that they expect and will receive well. Booth (1984), Brun *et al.* (1984), and Smith (1984) offer sound advice on making presentations.

References

Booth, V. (1993). "Communicating in Science: Writing a Scientific Paper and Speaking at Scientific Meetings," 2nd ed. Cambridge Univ. Press: Cambridge, UK.

Brun, L. J., Bovard, R. W., Foss, J. E., and Stark, S. C. (1984). "Scientifically Speaking." SRI, USA.

Haakenson, R. (1975). How to handle the Q & A. *In* "A Guide for Better Technical Presentations" (R. M. Woelfle, Ed.), pp. 158–170. IEEE Press, New York.

Smith, T. C. (1984). "Making Successful Presentations: A Self-Teaching Guide." Wiley, New York.

14

COMMUNICATION WITHOUT WORDS

"It is the province of knowledge to speak,
and it is the privilege of wisdom to listen."

OLIVER WENDELL HOLMES

In any communication effort, we must recognize the transfer of information without or in addition to the words we use. Listening and reading constitute at least half of communication. In speech, every gesture and every facial expression add dimension to what we say. With written and visual media we also use language beyond the words. Technical editors are trained to use type styles and sizes as well as other symbols for emphasis to guide a reader through a paper. As we design manuscripts for photocopy or for publication from discs, we become responsible for symbolic language that used to be in the domain of editing. Symbols, spacing, colors, and other embellishments can be as important as words in written and visual communication. In speech, the physical circumstances, body language, and listening habits all contribute to success or failure of information exchange.

SYMBOLS

Conventional symbols are available and new ones can be designed to help us express our precise meanings. Conventional symbols include placement and size of headings that constitute guideposts to lead a reader from one section of a paper to another. Even more common is the paragraph indention. We have become accustomed to italics to denote scientific names of species and abbreviations to designate units of measure. We accept symbolic conventions often without even noticing them. Words themselves are actually symbols for sounds and meanings. We stop at a period as we stop at a red light without analyzing why. A great deal of simplicity in communication would be made complex if such symbols were not universally accepted in a language. The important thing to remember about your use of symbols is to be sure that your audience will interpret them as you intend them.

Dreyfuss' (1984) *Symbol Sourcebook* is based on a data bank of over 20,000 symbols. These symbols are a supplement to all languages. Some, such as the skull and crossbones image that indicates "poison," are recognized worldwide; others are limited to a small group of people or perhaps to specialists in a given academic discipline. Sign language, Morse code, chemical structures, Braille, traffic signs, and hundreds of other symbols all carry important messages.

Size, shape, spacing, color, and location as well as underlines and bold print can be used with discretion to help organize, emphasize, and clarify meanings in your writing or in visual displays for posters and slides. Without discretion, such symbols can simply confuse an audience. If every third word in a text is underlined, the underlining has no meaning beyond counting words, but if only one word in a paragraph is underlined, the reader interprets the symbol as suggesting emphasis. The use of large print, boldface, color, total capitals, white space, and positioning all call attention to a spot or a word. For example, a headnote set in 14-point type, in total capitals, in boldface or italics, centered or underlined over an indented paragraph in 10-point type indicates that the headnote is the subject of that paragraph. For such a note to follow the paragraph or to be buried within it not only would lack the intended meaning but would confuse us by interrupting the conventional arrangement of ideas.

All these points seem obvious, but prudent use of symbols can facilitate understanding, and confusion can originate from careless or imprecise use that might create false emphasis or ambiguity.For example, if you label items in a list as 1, 2, 3, and 4, the reader may assume that 1 is more important than 2. Designating items with a, b, c, and d may be less likely to draw that assumption, and using a consistent bullet or shape such as ○, ▶, or * in front of each item is even less likely to suggest that the first item is most important. However, for complete clarity, you may need to tell your audience that "the following items are of equal importance" or "are listed in order of importance."

The overall appearance of the page or visual aid conveys a message even before the audience has read a word. Increasing the widths of margins to place a section of text inside the main text indicates that you are quoting another source of information or presenting an example. Successive indentions may indicate less importance or a dependence of the material with the deeper indention on that above it. The same is true of smaller type placed against larger type, and thick lines are more emphatic than thin lines. The possibilities are almost limitless for combinations of spacing, underlining, boldface, and other elements of emphasis.

TYPE STYLES

Especially with word processing, slide making, and poster construction, you need to understand something about type sizes and styles. Printers have various styles of type that they refer to as "faces." We select a font that consists of a size and face of type. Faces have names such as Times Roman, Script, Old English, and many others. Size is measured in units called "points." What you are reading now is 10-point New Caledonia type; the title for this chapter on page 137 is in 22-point New Caledonia.

Another printer term that you may need to know is serif. A serif is an extension beyond the main body or shape of the letter. The small horizontal extension at the bottoms or tops of letters like l or h in New Caledonia are serifs. Type called sans serif is blocky without these tails or extensions.

This text is set in 12-point Helvetica.

Notice that, even with type 2 points larger, I get more words per line with this style. The difference occurs mainly because Helvetica is a proportionally spaced font whereas New Caledonia is a fixed-space font. With Helvetica, wide letters such as m or n take up more space than the i or l. In New Caledonia, each letter occupies a fixed and equal amount of space.

Be conscious of type styles and sizes when preparing a paper for publication or reproduction by photocopy. A scientific poster or word slide can also be much clearer if the text is in a large, easy-to-read font. *Keep in mind that lowercase letters with only the grammatically necessary capitals are easier and faster to read than total capitals.* See the examples of type sizes and styles in Chapter 17 on posters (Table 17-1).

COLOR

A great deal of variety can be accomplished without color. Color simply adds another dimension to communication. Dark type on a pale background (or black on white) generally evokes little response from the reader beyond the meanings of the words themselves. But insert a bright red word, and the symbolic meanings with color can elicit a complex and even an emotional response. As Imhof (1982) says, "the concept 'color' is ambiguous." Scientific analyses of color do not take into account the emotional response a viewer may have to color. Because of the ambiguity in reactions, one can hardly provide a set standard for communicating with color, yet it is an important element especially in scientific slide presentations and posters.

Imhof's (1982) theory of color provides a valuable guide for cartographers that we can apply to other scientific communications. He maintains that "A color in itself is neither beautiful nor ugly. It exists only in connection with the object or sense to which it belongs and only in interplay with its environment." For example, shiny red means one thing on an apple, it means another on a football jersey, and still another on an exam a teacher has just returned to you. Besides the effect of environment on meaning for color, one's own personality, past experience, or the culture in which he or she lives may give to a color a particular meaning in a given context. People call colors harmonious or clashing. We speak of cold colors and hot colors. Blue is cool or cold; yellow is warm or hot. There are hot pink and cool green, earth colors in shades of brown and orange, and neutral gray that some call drab and others beautiful. Imhof (1982) says that "grey is regarded in painting to be one of the prettiest, most important and most versatile of colors. Strong muted colors, mixed with grey, provide the best background for the colored theme." This remark from a color expert may provide an important message for us when we choose background colors for posters or slides.

Dreyfuss (1984) says that "Color produces immediate reaction and is the exclamation point of graphic symbols, so it must be reckoned with." It **does matter** what colors the scientist selects for communicating an idea. For color selection, we rely on the natural environment for some meanings and on established customs for others. Hot pink would seldom be selected over a cool blue to indicate a geographical body of water or even as a background color for scientific slides. We have established that for traffic lights red means stop, green means go, and yellow is a caution. When you use color to convey meaning, be sure to remember your color-blind colleagues. To them the top traffic light means stop and the bottom one means go. Spacing, size, or design may carry your message as well as color in such things as graphs and will not leave out this audience.

For slide making, today's film recorders are a godsend, but keep them under your control. Such slide makers can produce millions of variations in colors; such technology is far beyond the needs for simplicity in communication. Humans can hardly distinguish that many shades of color. Don't ever let color or any other ornamentation distract from the scientific message. Too many colors, large areas of strong or bright colors, or poor choice of color can be detrimental to communication.

As with the meanings of words, the scientist must select a color that is aesthetically pleasing but will not distract from the science by calling attention to itself. Ask for the opinion of reviewers before you make final decisions on colors to use. Imhof (1982) says, "Subdued colors are more pleasing than pure colors." He goes on to provide six rules applicable for map design that may be equally applicable for other scientific presentations (see Appendix 10). I strongly recommend that you read Imhof's chapter on "The Theory of Color," and when you are making graphs or other illustrations, consult Tufte (1983) and *Illustrating Science* (CBE, 1988) for details on communicating with color and designs.

PHYSICAL COMMUNICATION

In spoken communication, the physical setting in which you talk, your physical presence, and your body language can be as important as any words you say. If you have control over the **setting,** make it as comfortable as possible for you and your audience. If you have no control, be conscious of such things as lighting, noise, temperature, the arrangement of chairs, or other attractions and distractions. Sometimes you will have to make decisions between the lesser of two evils; for example, choosing between a comfortable temperature and a noisy furnace or fan is not easy. Your words are worthless if your audience cannot hear them or are too distracted by other elements in the environment.

The value of your words is also diminished if your own physical presence or your **body language** is too distracting. Every movement you make, small or large, contributes to your communication. You talk with your eyes, with your feet, with your posture. Your grooming is one of the first messages you present to your audience. It's idealistic to suppose that the audience will "pay attention to what I have to say, not how I look." What you should wear depends on what your audience expects. The way you style your hair is certainly your business, but it still conveys a message. Be attentive to your grooming. Styles do change, but conventions in clothing and appearance are relatively stable in the scientific community.

Bodily expressions are as important as grooming; many expressions are extensions of your personality. You do not change your personality for an audience, but you can condition much of your body language within the confines of your personality. One student had a habit of unwittingly batting his eyes as he spoke, and in making speeches for my scientific presentations class, the constant blinking was most distracting. We were able to call his attention to the habit, and with conscious effort he controlled it and was a much better speaker with that single change in body language. Probably nothing is more important than eye contact whether you are talking with a single individual or a large audience. But you can also say things with your feet or your hands, a shrug of the shoulders, or a wave of the arm.

You may inadvertently emit distracting vocal sounds that clutter your message. Grunts, "ums, ahs, and you knows," a constant clearing of your throat, sniffing, or other audible distractions can cloud your message. Some of us don't have the most beautiful voices in the world, and a physical condition such as swollen adenoids or a cold may make it essential that we clear our throats, sniffle, or blow, but control these things as much as possible.

Such control may not always be possible. Don't despair if you have a speech impediment or a physical condition that attracts attention to itself but is out of your control. One of the most successful people I know stutters—not just an occasional repeat of an explosive B, but repeated interruptions with stuttering over words or letters. He has not allowed this condition to interfere with a career that demands public speaking and group communication. When his voice hangs onto a letter, he simply allows the stutter to run its course, takes control again with no apologies to anyone, and moves on with his speech. Audiences respect him. The first-time listener may be momentarily taken aback, but in almost no time that person is listening to what is being said and, along with the speaker, is ignoring the stutter. If this speaker exhibited added nervousness or an embarrassing self-consciousness, the audience would also hang onto the impediment and lose some of the speech content. By example, the speaker simply guides the audience in how to react. If you have an uncontrollable voice problem, a physical deformity, or even a temporary bandage on your face or hands, you need not try to hide the appearance nor call attention to it by appearing embarrassed. Almost everyone has some sort of problem. Whether you are extremely beautiful, tall, short, thin, fat, or have to walk with hand crutches or sit in a wheel chair, an audience will notice. Accept that fact, and make your listeners as comfortable as possible by ignoring whatever it is that makes you different. Whenever you can, control your appearance as well as the physical situation.

Your body language can have a very positive effect on what you have to say. Voice inflections, volume, and tone can change the entire meaning of words or make a point more or less emphatic. A hand gesture, closed fist, open palm, or directional movement of the hand can help to clarify meaning. Facial expressions can indicate beyond the words whether you are serious or jesting and whether you believe in what you are saying. Good use of body language can make an audience believe you have said something important whether your words are meaningful or not. You should not mislead audiences to believe what you say is important when it is not, but spoken communication is made most effective with use of physical expressions to reinforce the words.

Space is also important in communication. How close you stand to an individual or to a audience conveys a message. Humans are territorial; to get too close may invade their space, especially with individuals. However, a step forward toward a group suggests bringing the message closer or including them more fully. This technique is often used when a speaker invites the audience to ask questions. On the other hand, to back up too far or to plant yourself too firmly behind a speaker's stand can give the impression of excluding the audience. Our convention of the speaker's stand or simply of the speaker standing in front of a seated audience is important in separating the speaker and listeners into their respective roles. We would not have the same respect for the speaker if he or she should be seated in a chair out in the middle of the audience. People in the immediate vicinity might be flattered and attentive, but those 10 rows, or even 2 rows, behind the speaker would feel left out even if the voice were clearly audible.

Your personality, cultural background, physical condition, needs, or beliefs can influence your mannerisms so that you are not conscious of your facial expressions or body movement. Ask an observer to point out the strengths and weaknesses in your body language. Control the expressions that enhance your scientific message as well as those that distract. Don't let your words get lost among physical distractions.

LISTENING

We often underrate the importance of listening. Perhaps half of the responsibility for spoken communication should be that of the listener, yet we design full courses for teaching speech and give little or no attention to teaching people to listen. Listening is a matter of hearing, observing, and thinking all at the same time in order to have the clearest perception possible of what a speaker is conveying. Most of us have bad listening habits, but they can improve with practice.

Perception, or comprehension, not just hearing, is the end objective. In fact, the ears are only a part of listening. The listener's eyes must watch the speaker and be sensitive to his or her physical expressions as well as the words voiced. And be sure your brain is engaged with the ears and eyes in order to avoid misinterpreting what is said. To listen well, one must be sensitive to all the elements in the communication. Pay attention.

It takes concentration and energy to listen well. As with other forms of communication, you have to maintain the right attitude. Many of us have expended a great deal of energy and concentration in our lives on appearing to listen when we are actually tuning things out. We had other things on our minds when parents or teachers were talking, but we developed an ability to hold the proper facial expression and look at the speaker while our minds wandered. Or rather than paying attention to what is really being said, our thoughts are planning a reply to what we think is being said. Just half listening or "listening with one ear" is as bad as mumbling or just half speaking. Listeners must do their part to make communication a success.

Listening in a small group in which ideas are being exchanged is probably the most difficult listening challenge. You must be able to change your focus quickly from one personality to another and synthesize or recognize relationships in the material from several points of view. Listening as a single receiver of information or as a member of an audience requires variations of the same good habits. A conscious effort on your part could lead to better listening to a single speaker or to group communication. Try the following:

1. Look at the speaker. If you need to take notes, don't let that action make you miss the nonverbal communication.

2. Be sensitive to the tone and inflection in the voice. Watch how body language and voice blend into meaning.

3. Don't interrupt. A speaker may be struggling for words that you could easily supply, but let him or her finish an idea before you inject a response.

4. Concentrate on the subject. Relate what the speaker is saying to what you already know and, thereby, increase your knowledge.

5. Tune out the distractions. No setting is perfect. There will be noise, poor lighting, something else on your mind, or another member of the audience making distracting sounds or movements. Focus on the speaker and not the distractions.

6. Watch your own attitude and try to empathize with the speaker. Be open-minded. Don't jump to conclusions or judge the speaker by his or her appearance or accent, and don't completely evaluate what is being said until you have heard the entire speech.

7. Interrupt with a question only if you are confused. When the speaker is finished, be sure you have understood meanings. Ask questions or paraphrase what was said to clarify any nebulous points.

8. Be sure to respond. In a small group or one-on-one conversation, you can use your facial expressions or other body language to indicate when or how you are interpreting what is being said. You can ask questions to lead the speaker toward clear explanation. And when the speaker is finished, you can make your comments or observations. In a large audience, you may think your response is not important, but it is. Even if you cannot ask questions or make comments, your attention, posture, and facial expressions are important to the speaker.

9. Be patient. You can listen much faster than the speaker can talk. Use that extra time to think about what the speaker is saying and to make your nonverbal response. Don't be distracted. It is demoralizing

to a speaker to say, "Go ahead, I'm listening" when you are obviously being more attentive to someone or something else.

Not just in visual displays but with any written or spoken communication, the value Imhof (1982) gives to "the composition as a whole" is fitting. Your manuscript or your speech is not just organized words and ideas. It also contains symbolic images whether they are created by type size and style, color, physical circumstances, or body language. Concentration by the listener and the reader are also tantamount to successful communications. Pay close attention to your words and how they are organized, but also be conscious of the appearances and the listening habits of both you and your audience.

References

Council of Biology Editors (CBE) (1988). "Illustrating Science: Standards for Publication." CBE, Bethesda, MD.

Dreyfuss, H. (1984). "Symbol Sourcebook—An Authoritive Guide to International Graphic Symbols." Van Nostrand Reinhold, New York.

Imhof, E. (1982). "Cartographic Relief Presentations." de Gruyter, New York.

Tufte, E. R. (1983). "The Visual Display of Quantitative Information." Graphics Press, Cheshire, CT.

15
VISUAL AIDS TO COMMUNICATION

"Blessed is the man, who, having nothing to say, abstains from giving us wordy evidence of the fact."

GEORGE ELIOT

As with other symbolic images, visual aids can complement or supplement words in both written and spoken texts. Their basic functions are to attract attention to and clarify points beyond what the spoken or written word alone can do. In scientific communication if they attract attention only to themselves and do not increase clarity, they serve as distractions. Used effectively, they will strengthen the scientific message. Attention to the following five principles will help you avoid pitfalls in the use of all kinds of visual aids:

1. Make them simple enough to be comprehended easily.

2. Make the images or letters large enough to be seen clearly.

3. Make a trial run long enough before the real presentation to permit you to change the visual aids if they do not serve you properly.

4. Coordinate them carefully with the speech or text so that the audience is not distracted and the visual aid is relevant to the point in the speech or written text.

5. Just before a presentation that incorporates visual aids, be sure that any needed equipment is in place and operating smoothly.

Visual aids can add information, they can illustrate or provide examples or evidence, or they may repeat what you are saying or writing. In the poster, you can present information via a photograph, and the written text will just call attention to that information. Similarly, in a slide presentation you may describe in words the effect of a chemical compound on a petunia plant. Your words will be strengthened if you have a picture of a treated petunia to illustrate your point. In writing you have been taught not to be wordy or redundant. In speaking, repetition is often the best way to strengthen a point. A reader can return to a point in a text and read it again. Because a listening audience does not have the option to listen again, a judicious speaker will often reinforce with repetition. Visual aids can help you duplicate a point in a second medium without appearing redundant.

Visual aids can also help compensate for language barriers. When your first language or your accent is not the same as that of your audience, the importance of supplemental aid increases. It is especially important that you display key words you have trouble pronouncing or those that are new to the audience. With this assistance from your visual aids, if you will speak slowly and enunciate carefully, an attentive audience will understand you. I am not just referring to international students or to professionals from non-English speaking countries who join your societies. The English-speaking world includes numerous accents. Whether you are from Burundi, Bangkok, Boston, Baton Rouge, or Bath, your accent will be foreign to someone in your audience. Along with being proud of the way you sound, be sure you provide any assistance that your audience may need for understanding you. The best advice I can give is **SLOW DOWN** and provide informative visual aids.

In recent years, slides have been the dominant visual aids used at professional scientific meetings, but visual communication in posters may be as popular today. Traditional forms like transparencies still have a place in professional presentations and should be constructed with the same care that should be given to posters and slides. Choose the visual display that is effective for a given audience or situation. These include chalkboards, flip charts, videotapes, overhead transparencies, slides, posters, and other props.

Each of these kinds of exhibits requires its own equipment whether it be chalk and a chalkboard, a projector, or a poster board. Recognize the values and limitations of each. Flip charts and posters accommodate small audiences and require well-lighted conditions. Transparencies and especially slides require subdued lighting but can serve a rather large audience. Speakers often fail to note how handy a chalkboard can be. You should not write lengthy messages on it because your back is turned to the audience, but without other visual aids or in answering a question, it is often helpful to quickly sketch a chemical structure or write an unusual word for the audience to view. Videotapes or filmstrips interrupt the speaker entirely but can be coordinated with a speech if they are relevant. I attended a seminar once wherein the speaker introduced his subject, which had to do with the processes involved in germination and seedling emergence. After that introduction, with equipment ready to go at the push of a button, he showed a short time-lapse filmstrip of a soybean seed developing into an emerged seedling. He then provided a transitional remark and began his presentation of slides. The filmstrip was quite effective. But such interruptions can be distracting; be sure they fit smoothly into your presentation. Objects for show-and-tell can also be valuable or distracting. Any such prop needs to be big enough for all the audience to see and yet not ostentatious enough to distract throughout a presentation. Generally, passing an object around the room is also distracting; it may be better to invite the audience to stop by to view the display item after the speech. The important point with any visual device is to make it serve, not obstruct, the communication. Any visual aid should meet the following criteria:

1. Simplicity—Usually one point with limited subpoints
2. Legibility—Visible and legible to any person in the audience
3. Unity—Cohesive and uniform with other visual aids and with the written or spoken words
4. Quality—Clear, attractive, and aesthetically pleasing
5. Feasibility—Production and presentation possible with the materials, facilities, and time available

As with other symbolic communication, the style, layout, color, type, and form for a visual image should be clear and aesthetically pleasing. Characteristics of symbolic media may be as important as content in carrying messages to the audience. For example, color in a slide set or a poster can be a point of unity. Choose basic colors with care, limit the number of colors used, and coordinate them carefully with each other and with the content. Subjects, ideas, or transitions can be coded by color. For example, if you use a series of bar charts and, in the first, copper sulfate is represented with a blue bar, then copper sulfate should be represented with a blue bar in other charts used. In addition to color, other characteristics essential to the success of a visual aid include type, consistency, size, form, density, spacing, and style. Be creative but also use conventions that the audience expects. Computers can provide many advantages in creating visual aids, but they can also produce complexities or outright errors. Use their capabilities carefully. In this and other chapters, I discuss primarily visual displays used in slide or poster presentations, but the same qualities are important for any visual aid you use.

SLIDE COMPOSITION

In a slide presentation, along with a speaker and the speech, the slides carry the communication. Quality must be good to offer the proper support to the speaker. When you begin to make your slides, start in time to remake them if you need to improve the content or the quality.

With few exceptions, the scientific slide presentation will consist of photographs, word slides, and data display in figures and tables. A mix of these kinds of slides is more interesting than all of one. Photographs can provide information, transitions, and visual relief in a scientific slide set that has a great many word slides or tables and figures. As you pursue your research, keep a camera handy. You will one day need pictures of equipment, plants, animals, symptoms, or other images from your study. Microphotographs may be the most important evidence in the results of a study on a science that is invisible to the naked eye. For a slide or poster presentation, color photos are usually best, but also keep in mind that you may wish to publish a picture with a journal article and that journals often publish only in black and white.

Courses in photography and numerous books for beginners are available. Camera and film companies publish booklets and pamphlets that offer sound advice for photography. Plotnik's (1982) chapter on "Basic photography for editors" has good information for any photographer. Read such instructions, but as with other skills, you need experience to become adept. If you have not had extensive experience with photography, acquire a camera and give yourself a short course by reading instructions packaged with the camera and with the various kinds of film, ask questions of people at camera shops, and above all, shoot a few rolls of film. As you take photographs, visualize the final product; the image is in your viewfinder. Take several shots of the same subject with slightly different settings to the camera. Use slide film for slides and print film for prints. One can be converted to the other but never with the same high quality.

Be sure that the subject you want to display dominates the picture; some people take pictures too far away from their subject and include too much background. Certainly background is important. The photograph will appear two-dimensional, and tree limbs may look as though they are growing from people's heads. Placing your subject against a solid background, the sky, the earth, or an improvised backdrop can help. Lighting is important with both outdoor and indoor photography. Outside, the sun may be too bright, or it can cast heavy shadows that will subdue your subject. Watch the depth of field, and be sure your subject is the area in focus.

Many of your slides will be text, e.g., title slides, key words, and data in tables and figures. Whether you are producing slides with a film recorder or by hand, you must follow basic criteria of composition for good slide quality. The viewfinder of the camera and your resulting slide have relative dimensions of 2 units by 3 units. Try to keep the orientation horizontal; that is, the longer dimension will run across the screen. Fill the screen with select words or symbols that are spaced appropriately and are clearly visible from anywhere in a large room when projected onto the screen.

Plan your slides for viewing in a large room, and you can use them in a small area as well. A general rule of thumb is that a slide that is clearly legible without projection at arm's length (ca. 25 in or 0.65 m) can be read with projection onto a screen at a relatively long distance (ca. 50 ft or 15 m). To adhere to this general standard, make the text on the slide no longer than 14 lines (10 or fewer is better). Lines should never be longer than 40 spaces (34 or fewer is better). Note the following:

Objectives

1. To compose attractive, legible visual aids
2. To conserve both time and cost in slide production

This example is composed of nine lines vertically, and the longest line contains 25 characters horizontally. It will fit the 2 by 3 format for slides, and the 16 words are not too many for an audience to comprehend quickly, especially if you direct their attention to the slide with a pointer as you introduce each objective. More words or lines would make the slide heavy. Fill the screen but not with too many words or ideas.

In addition to spacing and sizes, in slide making be careful in combining colors or using a colored text on a colored background. For instance, a bright yellow may bleed into a blue background or cause a blurry whitish image at the interface. Use solid distinctive colors for such things as bars in bar charts but choose them carefully so that one bright color does not appear more important than others. Make trial runs with various colors of bars and lines to give you model slides to base your selection on.

These principles on size, spacing, and color are especially true for tables and figures. For a slide presentation on research, your data are the most important support to your objectives and conclusions. Present only the data points that are essential to your talk. More than 20 items in the field of a table are too many;

fewer than 16 is better. In a line graph, three lines are much better than five; and more than five are too many. In a bar chart, grouping can affect the number used. In groups of threes, nine bars are understandable; in groups of twos, eight bars are acceptable, but as single, separate bars, try to limit the number to no more than six in a chart. Let exceptions to these standards be rare. Often you can reduce the number of data points in tables and figures by recognizing that representative data will illustrate your points. If three different lines in a line graph run along similar data points, take two lines out and tell your audience that the data omitted are similar to those represented by the line on the slide.

Notice that Table 15-1 should not be made into a slide for two reasons: (1) the type would be too small for an audience to read, and (2) if they could read it, it contains too many numbers for them to comprehend in the short time a slide should remain on the screen. To produce an acceptable slide, choose representative data that best illustrate the point you wish to make. In Table 15-1 let us assume that you wish to emphasize differences in the total nitrogen. Note how Table 15-2 accomplishes this purpose and preserves an acceptable type size with the limited number of data points. As a speaker, you can point out any differences that you need to discuss relative to shoots and roots, or if you have time for the specific discussion on those points, you might produce two additional slides with that information. Three slides with limited information communicate better than one that is too heavy. The speaker also can provide the additional information that was in the original caption and footnote. Notice that the same limited data can be displayed in graphic form as in Fig. 15-1.

Some equipment allows for more flexibility than other, but you, not your equipment, are responsible for your slides. Make the equipment work for you. Note the following suggestions for composing word slides as well as tables and figures:

1. Use a thick, blocky type. Italics, script, or thin letters are harder to read.

2. Use lowercase letters, capitalizing only where most texts would capitalize. Lowercase or the mix is easier to read than total capitals.

3. Remember that projection will expand the spaces on the slide as well as enlarge the letters. Usually single space between lines of continuous text and double space only between ideas or to set apart a caption or heading.

4. When making word slides by hand on the light stand, be sure your copy is a strong contrast of black and white without smudges. Anything that shows on your copy will show as much or more on your slide.

5. Ordinarily, don't number your tables or figures. You may want to rearrange your talk and put Table 3 before Table l. With slides in sequence with the talk, the numbers are superfluous.

TABLE 15–1

Table 4 from Page 40 in the Thesis of David Mersky (1992), University of Arkansas, Fayetteville (Used with Permission of the Author)

Table 4. Influence of *Bradyrhizobium japonicum* USDA 110 (BR) and *Heterodera glycines* Race 3 (SCN) on Root, Shoot, and Total Nitrogen Contents of 'Lee 74' and 'Centennial' Soybean Cultivars.

Soybean cultivar	Treatment	Root nitrogen		Shoot nitrogen		Total nitrogen	
		%	mg/plant	%	mg/plant	%	mg/plant
Lee 74	Control	1.84 a*	11.5 c	2.46 bc	34.5 c	2.26 ab	46.0 c
Lee 74	BR	1.85 a	16.5 abc	2.94 ab	85.9 ab	2.68 ab	102.4 ab
Lee 74	SCN	1.78 a	14.3 bc	3.49 a	78.5 abc	3.04 a	92.8 abc
Lee 74	BR + SCN	2.03 a	21.0 ab	3.17 ab	96.7 a	2.86 a	117.7 a
Centennial	Control	1.57 a	14.4 bc	1.95 c	38.7 bc	1.82 b	53.2 bc
Centennial	BR	2.12 a	17.9 abc	2.55 abc	64.6 abc	2.44 ab	82.5 abc
Centennial	SCN	1.67 a	17.7 abc	2.52 abc	65.2 abc	2.28 ab	82.8 abc
Centennial	BR + SCN	1.83 a	23.6 a	2.69 abc	87.8 a	2.44 ab	111.4 a
LSD (P = 0.05)		0.75	7.1	0.97	48.3	0.88	53.8
CV (%)		23.59	24.0	20.63	40.4	20.59	36.1

*Means within a column followed by the same letter are not significantly different at the 5% probability level. Data are means of 3 replications.

TABLE 15-2
Adaptation from Mersky (1992, p. 40, Table 4) with Permission

Influence of *B. japonicum* USDA 110 (BR) and *H. glycines* Race 3 (SCN) on Nitrogen in 'Lee 74' and 'Centennial' Soybean[a]

Treatment	Total nitrogen (mg/plant)	
	Lee 74	Centennial
Control	46.0 c	53.2 bc
BR	102.4 ab	82.5 abc
SCN	92.8 abc	82.8 abc
BR + SCN	117.7 a	111.4 a

6. In typing tables, space evenly between columns. Pick out your longest line and space it to fill the screen horizontally. Then align other material with that. Always align decimals.
7. Be sure lines and symbols in figures adhere to the same standards as text. Thin lines or symbols that are too much alike may be difficult to follow when the copy is photographed.
8. Be consistent with abbreviations and other symbols from slide to slide in the same presentation. This consistency should include color,

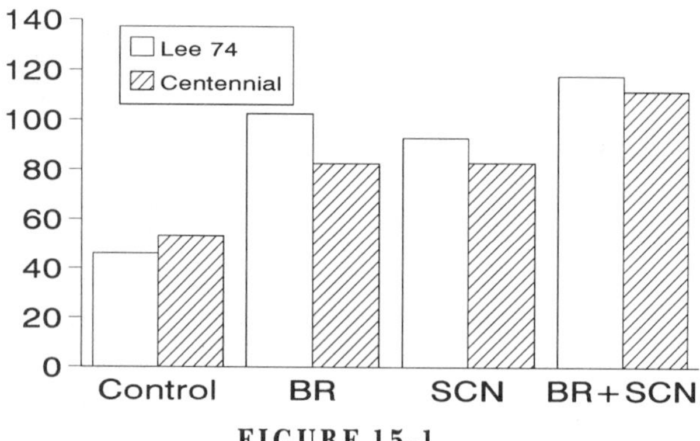

FIGURE 15-1
Graphic display appropriate for a slide.

symbols, axes, or any other communication device that is intended to carry the same message in one slide as in another.

9. Avoid the use of too many different colors. Any device for enhancement (italics, bold print, or underlining) that attracts attention to itself but serves no additional purpose is a distraction from the content.

10. Ask a colleague to review your slides and offer an opinion. What appeals to you may not appeal to someone else. You may want a second or third opinion, but if a point is questionable, there is probably a better way to compose the slide.

11. Make your slide set long before you need it and be willing to remake it.

MAKING WORD SLIDES

A number of techniques are available for making word slides. One method is not necessarily better than another; success depends, first and foremost, on your slide composition. No technique or film will make a good slide from poorly composed copy. Beyond the copy composition, the effectiveness of your slides depends on your judgment and your skill in using a chosen process.

For slides photographed manually with a camera on a light stand, be sure to fill the viewfinder as you would fill the slide screen. Your camera (35 mm) should be equipped with a macro (50 mm) lens for close-up shots. Level the camera and be sure the copy is parallel to the camera and in focus. It's good to use a cable release to keep from shaking the camera. Several kinds of film can be used. Kodak Vericolor can provide slides with various color selections for backgrounds (Loynachan, 1986). Depending on your equipment, you can also produce quality slides with high-contrast black-and-white film embellished with a colored transparency (gel) or a reversal film that creates a positive rather than a negative image. Dyes or even food color can be used to color the lettering on black-and-white slide negatives. Many of these ideas are not used much since the advent of film recorders and graphics software, but where equipment is limited, don't despair. You can still make good slides. Stock (1985) provides information on principles involved in making your own slides from typed copy on a light stand. Check other literature from the 1970s or 1980s (e.g., Loynachan, 1986, or Kodak Publication No. M3–106, Eastman Kodak Company, Rochester, NY 14650) or talk with your local film processor.

Computers have made slide making easier. Slides can be photographed directly from the screen, but take precautions to achieve clear copy. A monitor with a flat screen does not require compensation for curvature, but be sure the type size and style are appropriate and use a tripod to avoid shaking the camera. Experiment with lighting and film speeds. If possible, instead of photographing from the screen, use a film recorder. The film recorder produces the images electronically from software, and some very good software programs are now available. Try to choose a program and film recorder that can provide gradients in type sizes as well as several type styles, colors, symbols, and other embellishments that will allow you to follow the principles of good slide output.

Whether you photograph slides manually or through a film recorder, keep in mind the standards for slide composition. Although the computer can produce multiple boxes, colors, shadows, and intricate details, **a simple slide is still the best slide**. Computer clip art can help or hurt your slide. A neat unobtrusive logo or a simple bullet symbol can add to the communication, but don't get too cute with the artwork; it can clutter up a perfectly good slide. You want your audience to focus on the scientific message on the slide, comprehend that idea in a few seconds, and keep their attention on what you are saying. Don't let that focus be distracted by something clever that your computer program will do.

Whatever method you use, you may wish to run a trial strip of film to acquaint yourself with how to get the best results with the equipment. Numerous films, cameras, computer programs, and other equipment are available. Check to see what your department or your company has and how to use it, or consult with a photography supply store.

Slides and other visual aids can contribute to the success of an oral presentation. *But better no visual aid than a bad visual aid.* At almost every professional meeting I've attended, I have witnessed some speaker who projected a totally illegible slide on the screen and told his or her audience, "I'm sorry this slide is bad, but here you see. . . ." Then the speaker would proceed to point to something I could not see. Please don't apologize for a bad visual aid; don't use a bad visual aid.

References

Loynachan, T. E. (1986). Use of Vericolor 5072 film for reversed text transparencies. *J. Agron. Educ.* **15,** 43–46.

Plotnik, A. (1982). "The Elements of Editing: A Modern Guide for Editors and Journalists," pp. 121–143. MacMillan, New York.

Stock, M. (1985). "A Practical Guide to Graduate Research." McGraw–Hill, New York.

16

THE SLIDE PRESENTATION

"To speak much is one thing, to speak well another."

SOPHOCLES

You have done scientific research on a certain subject and have established a degree of expertise in that area. You are ready to make a slide presentation in your departmental seminar, at a professional meeting, at a job interview, or to a local civic club. My first suggestion is that you read Appendix 11, "Let There Be Stoning" by Jay Lehr (1985). He suggests that boring speakers waste our time and that they should be punished. We should demand excellence in professional presentations the same as we do in published papers. After reading Lehr's vivid analysis of what it takes to make a good slide presentation and before you start designing slides, think about who makes up your audience and what you want to tell them. With the audience foremost in your mind, consider your purpose and subject, your own personality and ability, the time you have, and any other influence on the outcome of your presentation such as the physical setting, other speeches, and the presence or absence of a moderator. Then prepare your presentation.

You may wish to write a draft of a speech, but don't memorize it nor read it to your audience. Many people can work better from a detailed outline than from a fully written speech, but don't rely totally on the written word in any form. Let your slides or a brief outline with key words guide you. Make the speech in conversational tones with as much eye contact with the audience and as little reference to notes as possible. Most important—condition yourself, allow time, construct a slide set you are proud of, organize material carefully, and practice with a reviewer.

FIGURE 16-1
Prepare yourself for the intensity you will feel when you begin to make the speech.

CONDITIONING YOURSELF

Before any oral presentation, you need to prepare yourself for the intensity that you should feel while you are speaking (Fig. 16-1). You need to practice, yes, but you also need to be relaxed and enthusiastic about your speech. If you fail to get enough sleep or if you practice repeatedly up to the last hour before the talk until the speech is almost memorized, you can sound tired and bored with the entire presentation. It is far better to prepare the presentation several days ahead of time, have someone review it, revise it, and then put it aside until just before you go on stage. Get a good night's sleep, dress appropriately for the event, and approach the talk with confidence and enthusiasm.

If you have questions about the organization of your report or if there are technical points that are not clear to you, spend some time before the presentation talking with your major professor or other colleagues who can help you. If you do not understand your subject, you cannot expect your audience to understand it after listening to you.

At many professional meetings, it is important to publish or hand out copies of an abstract before you begin your presentation. A well-written abstract can be a great advantage to you because the audience is listening for what you have to say about the major points in that text. An abstract must be very short or its purpose will be defeated. If the abstract is a handout for the meeting, a list of pertinent references can accompany it, but the abstract itself should contain no literature citations.

The abstract is an impersonal introduction to your subject. A moderator may give a little more personal introduction. Then the floor is yours for the designated time. If you are nervous before and during your presentation, do not be unduly disturbed. Most people feel some nervousness in the speaking situation; use that feeling to make you alert, eager, and animated. Confidence will soon take over if you are well prepared. Above all, don't let the nervous energy make you talk too fast to be understood.

Try to enunciate clearly. Make a special effort to speak slowly and loud enough so that those in the back of the room can understand you. Face your audience as much as possible. Look at them; eye contact is important for holding their attention. Try to avoid such nervous mannerisms as hiding your hands in your pockets, fiddling with objects in your hands, or unnecessary waving of a pointer; these displays are extremely distracting to the audience.

The pointer, the remote control for slide advance, and a microphone are equipment you must deal with while you are speaking. Practice with various kinds of these tools. You may be issued a light pointer or a stick pointer. Learn to be comfortable with either. A pointer can be helpful to direct and hold the audience's attention to a point on the screen, but turn it off or lay it aside when you are not using it. Otherwise, the light pointer can turn into a distracting series of flashes across the audience, and the stick can appear to be a baton and you an orchestra conductor.

The remote control may be connected to the projector with a cord, or it may operate with a wireless electronic sensing control. If the cord is attached, keep it to one side so that you don't become entangled. If you are bouncing the signal off a wall with a wireless control, be careful not to hit thin partitions between two meeting rooms. You may be advancing the slides of the speaker next door. That situation is most disconcerting.

The microphone is a valuable accessory especially in a large room or for a voice that does not project well. Take advantage of a good mike; it can help you provide quality in tone and enthusiasm. As with the pointer, you need to become comfortable with the mike. You may have to adjust to the kind used, and more problems occur for women than for men. The mike may be a clip-on that can be attached easily to a tie or suit lapel, but it can be unduly heavy on a thin blouse or dress and even worse if the garment has no lapel. The microphone may be stationary on the podium, or it may be on a cord that is put around your neck. These can be better for women, but anyone needs to be conscious of the range of the stationary mike and not walk too far away from it. You may even have a mike that uses a pocket remote sensor to a speaker system. These are simple devices that slip easily into the pocket of a man's coat or pants but may not be suitable for some women's clothing. As microphone systems become perfected and manufacturers recognize the need, perhaps women will not have to adjust their attire to fit the equipment.

TIMING

Recognize that timing is important during preparation and delivery of any talk. Boil down your remarks so that you can present the significant points and auxiliary explanation within the allotted time and give adequate coverage to each point of emphasis. As McCown (1981) suggests with posters, a good idea is to start with your abstract and enlarge upon the points it contains with examples and details. Planning for a speech and designing slides should begin at least 4 to 6 weeks before your presentation. In fact, as you do your research and data analysis, you should be planning for the delivery of results to an audience. Make slides and other photographs throughout your study. Even with modern technology, you should plan to have your presentation in its final form at least a week before the actual delivery. In addition to conditioning you to make a fresh and enthusiastic presentation, this time allows for the possibility of correcting errors in slides or improving poor communication efforts after a review by your peers.

You will seldom have an opportunity to set your own time limit for a presentation. For some situations, such as a job interview seminar, you may be given a generous range of time. A prospective employer may test your judgment by suggesting that you take "20 or 30, no more than 40 minutes." Wisely plan that presentation at 25 minutes; your audience will probably want to ask questions, and the attention span for most people is less than 30 minutes. In other situations you will be given a designated time within a program. That time may be very specific—"15 minutes, use 12 for the speech and 3 for questions." Such a request gives you very little leeway, but you will note that others have the same limits, and it is rude and unprofessional to finish too soon or to run over into someone else's time. The program chair expects that you will fill your time slot and doesn't want to get off schedule by starting the next presentation too early or too late.

Timing is also critical during the speech itself. Give your main point the most time. Don't spend much time on side issues. Usually the time allotted will allow you to make only one or possibly two main points. The way you use the time for exhibiting that point depends on your organization. The pace with which you move through the presentation is critical to audience reception. Slow down and remind yourself to provide the audience with time to think by using pauses, full stops, and a relatively slow pace. A friend of mine, who is an outstanding speaker, writes **SLOW DOWN** as a reminder at the top of each note card. If you have a tendency to talk fast, you might do the same. The speaker who gives the audience 2 or 3 seconds to think just after he or she makes an important point usually maintains the audience's attention far better than a speaker who drones on without clear punctuation in the delivery.

THE SLIDES

Even with the same subject, slides selected should be somewhat different for any given audience. Your colleagues in science may understand a complex graph that would be lost on the audience at the job interview or the civic club. The latter audience might enjoy seeing a picture of equipment you have used, but the scientists may have seen such instruments so often that, if you simply name the piece of equipment, an image is in their minds without the need for a photo. After you have thought about your audience and your purpose, only then are you ready to consider the content for slides.

Seldom should you put words on a screen and then say those exact words without further comment. In other words, don't read your slides to an audience. Exceptions to this idea are your title, your objectives, and your conclusions. Again, the emphasis with visual and verbal duplication is not only acceptable but can be highly effective. Be selective in what you duplicate. You may not wish to read the title verbatim, but it is usually best to read the objectives and conclusions and add commentary with each. For almost every other part of your speech, do not bore your audience by putting the text of your speech on the screen. Instead, select key words and other images, display them, and expand upon their meanings in your talk. This technique allows you to keep text on the slides that the audience can comprehend at a glance and then move their eyes back to you to get the most from your voice and body language. Use a pointer discreetly to direct or hold their attention to a point on a slide but lay that pointer aside when you are not using it. Study the principles for creating and using any visual aids before making your slides.

COORDINATING THE SLIDES AND THE SPEECH

As you organize your talk, consider the audience and the time you can give to each portion of the content. Speech teachers may tell you that you should roughly divide your talk into 10% introduction, 80% main body of the speech, and 10% conclusion. Often these percentages should be altered to provide more time at the beginning or the end for explanations that are crucial to a particular audience's understanding. In scientific communications, we may follow the classical beginning, middle, and end, but in research reports we are more likely to think of organization in terms of introduction, methods, results and discussion (IMRAD). Table 16-1 outlines typical contents for both the text and the slides for a research presentation. See the example in Appendix 12.

For any speech, with or without visuals, your introduction needs to let the audience know what your subject is and how you will approach it. Once you have identified your subject and provided a justification or hypothesis for the

TABLE 16–1
Typical Content for a Scientific Presentation

Speech Content	Slide Content[a]
I. Title	Title and authors
II. Introduction	
Hypothesis	Full statement
Justification	Key words, pictures
Literature	Ideas and references
Objectives	Full statements
III. Methods	
Equipment	Photos or illustrations
Sampling and technique	List or flow chart
Methods of analysis	Statement or key words
IV. Results and Discussion	
Objectives accomplished	Statement and pictures
Data	Tables, figures, discussion about data, key words, photographs
V. Conclusions	
Main outcomes	Full statements
Accuracy of hypothesis	Statement or photograph
Application of results	List or picture

[a]These are examples of forms that usually work well. Your content may require that you use other forms.

substance of your talk, the best introduction is usually a general outline of what will follow. "Today, I'd like to discuss two issues. . . . The first is. . . . , and the second . . . " or "We can follow the development of this theory through 3 stages: 1 . . . , 2 . . . , 3. . . ." The audience can follow you more easily if they have that initial outline of information. From that point, take the speech through the development of your points, and draw a conclusion by reiterating the main points and suggesting an interpretation based on your development of ideas. With any speech, decide what you want the audience to know and carefully set up your objectives and conclusions with adequate intervening material to support both. Then ask for a second opinion about how well you are communicating and revise the order if need be.

Paramount to making your slides aid you in your presentation is to have them well organized and carefully coordinated with your speech. *A title slide, an objective slide, and a conclusion slide* form the foundations around which you organize your talk and your other slides. Even these may be modified for different audiences, but they will exist in one form or another in any good slide presentation.

The **title slide** should focus on what you are going to talk about. It should be clear and aesthetically pleasing. Beyond your own appearance, it may be the first impression you make on your audience. In addition to the title itself, this slide should contain your name and the institution or agency you represent (if you do). Some people argue that, on most occasions, a host or a moderator will have introduced the speaker by name along with the title of the presentation and, therefore, the slide need not repeat that information. I believe such information bears repeating. Here is another legitimate use of duplication in speech. Although your title and your name have been announced, a visual aid can reinforce that information or provide it to those who did not hear the moderator. Especially at the beginning of your talk, some people in the audience are not paying attention. Your title and identity are important; double the listeners' chances of having that information.

The **objective slides** should briefly state the research objectives that you are going to discuss or else show the audience the purpose or outline of the speech you are presenting. Objectives should be limited in number; it is hardly feasible, even in a relatively long speech, to present more than three research objectives, and one or two are better. They should be worded as briefly as possible, but don't neglect clarity for brevity. You may wish to read each objective verbatim from the slide and then pause and discuss its meaning or why you have chosen that objective. Move slowly over the objectives; you may want to put each on a separate slide to slow you down. The objectives constitute the heart of your discourse; be sure to focus on them and give them the emphasis they deserve.

The importance of the **conclusion slides** also cannot be overstated. The listeners may forget most of what you have said, but if you present dynamic conclusions, they will be able to carry your point away with them. Remember that the content in your conclusions should be coordinated with that of the objectives. Your objectives should state your goals and the conclusions should indicate the extent to which you accomplished them. On the slides, conclusions can be in complete sentences, and as with the objectives, you need to move through them slowly and deliberately. Embellish each with reminders of what you have said in the body of the talk that led to the conclusion. Above all, keep you voice alive and enthusiastic. If someone's attention has wandered, it is time to demand that your point be heard. Use pleasant but forceful words and voice; punctuate with appropriate facial expressions and hand or body movements.

Once you have designed your title, objective, and conclusion slides, you are ready to fill in between them. Your first task along with the title is to make clear your point of emphasis, your subject, your hypothesis, and the justification for doing your study or for presenting your ideas. Even before your title slide, you may wish to show a photograph or two that provides justification for your study. For example, if you have been studying the habits of the American burying beetle, you may want to show a close-up of the beetle itself. Not many people have seen one in recent years. During the showing of no more than two or three such photographs, you can explain the value of the beetle in the environment, the fact that it is now endangered, and the need to understand its habitats and habits. Some of that justification can be done before the title or all of it after the title, but along with the objectives you will want to enlarge upon the justification as much as your introductory time will allow. Present your own hypothesis and reference to literature that relates directly to your study. For instance, you may need to compare or contrast your work concerning the American burying beetle with that on similar species. Objectives may conclude the introduction after you have justified your study, or they may come soon after the title prior to that justification. No one pattern for organization is better than another, except for a given subject and a given speaker. In other words, the **introduction** of your speech and the visual aids that accompany it will include a title and objectives and other supporting material that can be arranged in any of various patterns around these two essentials.

The remainder of your presentation will be much more lucid if this introduction is clear. Use the IMRAD formula for organization if possible, but do not impose it on an audience or a subject when it is not the best way to get your points across. Characteristically, you will move from the introduction to the **methods.** Think of what unspoken questions may be in the minds in your audience. They may be pertinent, or they could get you off track. Stick closely to your purpose. Your responsibility is to answer the right questions and guide the audience to your point. If you have been studying the American burying beetle, your audience may want to know how you trap the beetle, how you get a license even to handle the endangered species, what data you collected, under what circumstances you observed the beetle, what analysis you used—the list of questions can become endless. You must decide which questions are most relevant to the focus of your speech and how to answer them in the time you have to devote to methods. Let your visual aids then support your answers. You may have slides in the form of diagrams of the traps you use or photographs of the area where you trap beetles or of the traps themselves. You may have lists of materials, a statistical design, or categories of data such as eating habits or flight habits of the beetles. For an audience that is mostly interested in your results, do not spend too much time on methods unless your objective was to study methods. Your basic design for the experiment, the kinds of data collected, and the statistical analyses used usually provide enough information to tell an audience that you have pursued your objectives in a scientific manner; then go on to the results.

Again, your audience is crucial to the way you handle the **results and discussion** section. Present representative, selected data to illustrate what you have accomplished in pursuing your objectives. Discuss any special meanings and implications that can be drawn from the study. Fitting your study in with those of others researchers further increases your credibility. Make reference to specific literature and use the names of other researchers. Such specific references will enhance your reputation far more than a general allusion to "other researchers."

For those at a professional meeting who understand your technical and scientific terminology and are accustomed to seeing line graphs and bar charts as well as tables of information, you can rely heavily on these forms to present results. For another audience, you may need more word slides or photographs. Certainly, for any audience a mix of word slides with field and laboratory photographs and graphic depiction of data is more informative and pleasant than all words or all photographs. Precision and clarity in communication can be increased by using a variety of different techniques for presenting an important point. In your presentation on American burying beetles, you might show results by providing numbers in a table, then a bar chart of information, and then pictures of the beetle itself to illustrate your observations and discussion.

Hold to your point of emphasis throughout your presentation. As with the methods, results and discussion should flow from the objectives to establish whatever resolution your data provide for your hypothesis. Remind the audience of your objectives, and then show the results and discuss their significance relative to those objectives. Once you have clearly enumerated results and discussed them in relation to your main point of emphasis, you are ready for the conclusions. Be sure the conclusions clearly reflect the goals of your objectives.

Throughout the presentation, carefully coordinate the slide on the screen with the words you are speaking. Practice enough to be sure that the order of the slides is the order of the talk and that you are comfortable with both. Critical to a smooth flow of the speech from beginning to end are the transitions that punctuate the presentation.

TRANSITIONS IN A SLIDE PRESENTATION

Often, the difference in an adequate presentation and an excellent one results from the pace of the speaker and the transitions he or she uses to move the audience along. Transitions slow you down and give the audience time to think. Transitions get you into and out of the speech and move your content from one section to the next, from one point to another, and even from one slide to the next. The typical slide presentation based on a research project contains several strategic junctures at which you must carry the audience from one point to the next with clear transition.

The first transition many people overlook is the juncture that occurs as you begin the talk. If you have no one to introduce you, you will have to attract the audience's attention to you and your speech. Some in that audience are talking among themselves about subjects entirely unrelated to your speech, some are thinking about their schedules or their own speeches, and some are simply concerned with the weather or their comfort. Transitions here include such things as closing a door, stepping to the front of the audience, stepping forward toward the audience, dimming the lights, or saying something like, "Let me have your attention, please." Most professionals interpret all of these gestures as an indication that you are about to begin. You may have to pause as the talkers finish a sentence or a latecomer finds a seat.

Of all the transitional devices in the world, the *pause* is often the most effective. After you have made a remark, closed a door, and stepped forward toward the front row of your audience, a pause will do wonders to bring the audience to you. Even those in their own private worlds of thought will look up at a silence to ask themselves, "What's going on?" or "What's the speaker waiting for?" The key is to pause long enough to attract the attention but not long enough to lose it again.

A part of the responsibility for moving the attention of the audience to the speaker may belong to a moderator. This host may be the one to shut the door, to adjust the lights, to call for attention, and to make the strategic pause. Whether you or the moderator assumes these duties, the transition is not complete until you begin your speech. Once you have placed yourself in front of the audience and are ready to begin your speech, give the audience a moment to adjust to the way you look and sound. A smile, eye contact across the audience, and an introduction to your voice are in order. Thank the one who introduced you, and then let your first sentence lead the audience toward your point, but do not let it be critical to their understanding your main point. Their minds have a lot of baggage to check before they can join you. They have to put aside whatever else they were thinking and get used to you and the sound of you. Making this initial transition will probably take no more than 10 seconds, but it is crucial to a good beginning.

Equally important is the manner in which you leave the audience. Without a transition, an awkward moment occurs as the last word in your prepared speech is spoken and you are ready for questions or ready to leave the audience. A moderator can be valuable at this juncture also. If the host will ask for questions, readjust the lights, or dismiss the audience, you have the moment to let your mind relax, turn off your last slide, or prepare to answer questions. If you are alone, you must assume the responsibilities of both you and the moderator. At this final juncture, you must recognize how much time you have left, maintain a professional stance, and let the audience know what comes next. Do not loosen your tie or roll up your sleeves or stuff your hands in your pockets. If a question period is to follow, do not remove a microphone that has been attached to your lapel or around your neck. Be sure the last slide is turned off and the lights are turned on for the audience. Whether you or the moderator asks for questions, a good transitional gesture is to move a step or two toward the audience. This symbolic language suggests an invitation for the audience to join you; you are joining them. After you have answered the last question, the moderator may assume the transitional duties of making a closing remark, thanking you and thanking the audience for coming, and providing any additional instructions (e.g., there will be another speech, the management has asked that we exit to the right). Moderators are handy things to have. Without one, you need to summarize the talk and the questions, provide any instructions the audience needs, and thank them for listening to you. Above all, be sure to finish in the time allotted to you.

Perhaps you had not thought of the periods of opening and closing the speech as transitions. Treat these entrances and exits with respect. They may be the most important transitions you make. More likely, you think of moving from one section of your speech to another as needing transition. How do you leave the introduction and move into the methods, or how do you leave the discussion and move to the conclusions? These certainly are critical points at which you must carry the audience from thinking of one thing to thinking of another. Give the audience a smooth ride and they won't even notice the transition. In a slide presentation some speakers will use headnotes on slides to indicate a new section of the speech. "Introduction," "Methods," "Results," "Conclusions"—these signposts provide obvious directions for your listeners and are often appropriate and inoffensive. However, they can be awkward impositions that could be avoided by using a relevant photograph or speech alone. With or without the headnotes, your words need to carry the audience to your next section. To get into methods from the introduction, you may remark, "To carry out this study, we set up two experiments. The first. . . ." Or you may say, "To accomplish my objectives, I first acquired. . . ." These remarks are smoother than a blunt announcement of "materials and methods" and are just as effective. Similarly, as you move into the results and discussion, you may have some sort of transitional slide, but you should also let your words carry the transition. Try something like, "The data collected showed that our hypothesis was accurate" and then move to a results slide that supports that contention. Or you may need to say, "Our results are inconclusive, but they do show that. . . ."

Moving to the conclusions is critical. You must have the full attention of the audience in order to leave them with your main points. Sum up the essence of the results you have presented and move dynamically to the conclusions: "All of these results point to two conclusions" or "Although the data are limited, we can conclude that . . . " are simple and often effective entrances to the conclusions, but you may be able to be more creative than these remarks are. Let's go back to the American burying beetle. Your hypothesis may have been that the lights of the homes and cities across the country are attracting these beetles away from their natural habitats in dead carcasses and from their habit of setting up homes in these habitats to produce and feed their offspring. Your conclusions need to lead us back to that hypothesis and your objectives. You could move from your results to the conclusions by declaring that "Our results indicate that American burying beetles will not keep their minds on home and family when they see the lights up town. We conclude that. . . ." You are ready at that point to exhibit your first conclusion slide, which will most likely reflect the outcome of your first objective. Your remark has given the audience time to digest their last thoughts about the data you have just presented and to move with you to the conclusions. Remember the value of the pause. A deliberate pause with a transitional statement can effectively attract the audience's attention. Let the slide on the screen fit the occasion and serve to support the pause and the remark. For my burying beetle, a photograph of the beetle at a light or moving toward one would be appropriate.

In addition to transitions into and out of the speech and at major junctures, you need clear transitions all along the way. Transitional word slides or photographs can help. Whether a picture or words are on the screen, the important point is that the transitional slide is appropriate. Smooth movement from one idea to another may not allow time for an extra slide. As you linger momentarily over one slide, you draw a conclusion to that point and then move with a short transitional word or phrase to the slide displaying your next idea. For example, as you leave one data slide, you could be saying, "Contrary to what we found with these data . . . [advance to next slide] the analysis of. . . . showed that. . . ." Such transition should be woven throughout the presentation.

Be careful not to keep repeating weak transitional words or phrases that become worn thin with overuse. Some to use with great caution include, "Also, we see. . . . ," "And here you can see . . . ," "Again, . . . ," "Looking at the . . . , we see that. . . ." or nontransitional utterances such as "Uhh," "And uh," or "OK." As Booth (1993) suggests, "'Anderm' [is] the most irritating non-word ever misfangled." Make your transitions provide the audience with both meaning and think time. Instead of saying: "Again," give the audience a little time to move from one point to another with a transition such as, "Those data show us the anatomical differences, but other data [advance slide] illustrate that physiological differences are just as important." Yes, it takes longer to make that statement than to say "again," but the audience has time to think about anatomy and be ready for the next point on physiology. This think time is essential to any good communication.

THE PEER REVIEW

When you believe you have your content organized, laced together with transitions, and proportioned relative to time, go through your speech with your colleagues. Always accept healthy criticism. Their vantage point is closer to that of the audience than yours is. If slides or speech content and organization are not clear to these reviewers, revise your materials until they are. As you make more and more talks, you will become more sensitive to what works for you and the audience. When you have gained experience, don't be afraid to go it on your own, but always welcome reviews. In addition, during the speech be sensitive to the audience reception and be willing to deviate from your prepared material if you recognize that their attention or understanding is straying.

These general comments on timing, organization, and review will not make a good speaker of you. As with learning to swim, the only way to develop the skill is to do it. If you follow the basic conventions for good speech making and repeatedly practice the proper moves, you can become an accomplished speaker. Booth (1993), Brun *et al.* (1984), and Lehr (1985) have good suggestions for making scientific presentations. The following checklist may alert you to points to observe in your own scientific presentations and those of others.

CHECKLIST FOR PROFESSIONAL SLIDE PRESENTATIONS

I. The speech
   A. Introduction
      1. Are your hypothesis and objectives clear for the audience?
      2. Do you provide the audience with clear rationale and justification for your study?
      3. Does your introduction follow a logical pattern, and is it related to other literature and to scientific principles?
   B. Materials and methods
      1. Do your methods have the support of the literature and scientific principles?
      2. Do you show a logical, step-by-step process for executing the experiment and collecting the data to carry out your objectives?
      3. Do you make clear your use of appropriate experimental design and statistical analyses?

C. Results and discussion
   1. Do you summarize results, i.e., emphasize main points, as you *begin* and *end* this section?
   2. Do you relate the results clearly to your objectives?
   3. Do you carefully choose a limited number of data points to support your contentions and present them in simple illustrations, graphs, tables, and lists?
   4. Do you discuss your points in terms of
      a. Their relation to other research.
      b. Their practical or scientific applications?

D. Conclusions
   1. Do the conclusions reiterate main points for the audience to remember?
   2. Do you show a list and clearly relate it to your objectives?
   3. Do you give examples of application and use for your findings?

II. Visual aids

A. Number—Are there too many or too few slides for the time you have?

B. Content
   1. Are the slides clearly coordinated with your speech?
   2. Is the purpose of each slide readily apparent?
   3. Do you have a balance of data, lists, and information slides with photographs interspersed throughout?
   4. Have you included all expected slides: title, list of objectives, and conclusions?

C. Quality—Are your slides
   1. Neat and spaced to fill the screen?
   2. Simple and free from excessive data?
   3. Easy to comprehend, e.g., proper size print, good content, good design, clearly labeled axes?
   4. Free from garish color or any other embellishment that could distract from your message?

III. Speaker

A. Are you prepared?
   1. Are you familiar with your speech and the slides?
   2. Will you and your audience be comfortable with your appearance?

B. To what extent do the following support or distract:
   1. Mannerisms and gestures?
   2. Audience contact (eye contact, and facial expressions)?
   3. Voice, speech patterns, and ease in speaking?
   4. Your attire, posture, and poise?

C. During the speech, keep the following in mind:
   1. Avoid reading from the slides or from notes.
   2. Be sure your eye contact covers all the audience.
   3. Put the pointer down when you're not using it.
   4. Don't put your hands in your pockets.
   5. Never apologize or make excuses for a bad slide. A bad slide is worse than no slide.
   6. Keep your voice enthusiastic and loud enough.
   7. Be sure to use enough but not too much time.

Recognize all the elements that come together in a successful slide presentation, practice them, and be proud of your performance. A good slide presentation coordinates communication by the speaker, the speech, and the visual aids. It has a specific purpose and is directed toward a specific audience. Be conscious of all the influences on the speaker, the speech, and the visual aids. Check everything including the physical setting, the pointer, and the microphone. Prepare with good slides, clear organization with appropriate transitions, and a peer review of the presentation. During the talk, be alert to the reception by the audience and the role of the moderator. Maintain a professional attitude before and after the talk as well as throughout the presentation, which includes the question–answer session. Confidence and a dash of humility will lead you to success.

References

Booth, V. (1993). "Communicating in Science: Writing a Scientific Paper and Speaking at Scientific Meeting," 2nd ed. Cambridge Univ. Press, Cambridge, UK.

Brun, L. J., Bovard, R. W., Foss, J. E., and Stark, S. C. (1984). "Scientifically Speaking." SRI, USA.

Lehr, J. H. (1985). Let there be stoning. *Ground Water* **23,** 162–165.

McCown, B. H. (1981). Guidelines for the preparation and presentation of posters at scientific meetings. *HortScience* **16,** 146–147.

17

POSTER PRESENTATIONS

"Only the composition as a whole determines the good or bad of a piece of graphic work."

EDUARD IMHOF

Posters have become a major format for communicating at scientific meetings. These displays of research findings use visual and verbal information with illustrations, the written text, and spoken explanation by an author. The technique varies with different societies, but generally the poster will be on display for several hours, perhaps all day, and the authors will be present during a part of that time to discuss the subject with viewers. Depending on the meeting, the number of posters displayed at one time may range from a dozen to several hundred. The audience is always a relatively small group of sincerely interested people. Presenting a poster is a good opportunity to build your reputation as a confident, knowledgeable, articulate scientist if you exhibit an attractive, informative display and maintain a professional demeanor as the author.

The professional poster session now seems to be a progeny of communication that just naturally evolved from other scientific presentations, but 20 years ago it was practically unheard of. Posters were introduced into scientific meetings in the United States in the mid-1970s (Maugh, 1974). They rapidly became a way to display large numbers of research efforts and have been widely accepted as a viable complement for and alternative to slide presentations and symposia or workshops. The Tri-Societies (American Society of Agronomy, Crop Science Society of America, and Soil Science Society of America) first used the technique at their national meetings in 1977 when 50 to 60 posters were presented. By 1990 that number had increased to 1215, almost 50% of the total presentations, and it has held at about 50% since then. Other societies have similarly increased the use of posters. Its rapid acceptance underlines the advantages of this format.

Posters offer advantages both for meeting arrangements and for communication efficiency. More papers can be scheduled for the same time with posters than with oral presentations, and those attending meetings have access to more papers in the same amount of time. Some convention centers can now provide large areas for display more easily than they can provide numerous meeting rooms and audiovisual equipment.

The advantages to individual communication are as appealing as the tactical convenience in arranging for a convention. The method provides a two-way interaction that is less feasible with the slide presentation. The poster presentation may turn into a very profitable question and answer session with both the presenter and the audience deriving mutual benefit from the ideas exchanged. Compared with the slide presentation, the poster technique provides more convenience in following up on ideas. Names and addresses or phone numbers can be easily exchanged, and the scientist will be more likely to contact the person he or she has spoken with face to face.

Imhof (1982) declared that "Only the composition as a whole determines the good or bad of a piece of graphic work." What's true for Imhof's maps is also true for posters. A successful poster must communicate through every visual and verbal detail. Because posters are essentially a cross between the visual and spoken communication used in a slide presentation plus the textual substance of a written paper, the same communication devices can be adapted to the poster format. Posters perhaps balance the visual, oral, and written elements more fully than either the slide presentation or the written report. The basic communication devices used in all three include the text, the type size and style, color and texture, shape and arrangement, and illustrations of data in tables, figures, or photographs. In getting involved with all of these media, don't lose sight of the concept of unity or "the composition as a whole," and remind yourself of the basic purpose of any scientific communication: to convey a scientific message to an interested audience.

AUDIENCE

As with any other communication, with posters you need to have as much concern for the audience as you do for your subject and your own presentation of it. These viewers are 1 or 2 meters away as they read the poster (Fig 17-1). The material must be attractive, interesting, and clearly legible to keep their attention. Remember that the meetings offer many possibilities, and your

FIGURE 17-1
Have concern for the poster audience.

viewers have plenty of other things to do. Also, most readers are standing, and it is tiring to read from the standing position for very long. Most of them will look at the main points of your poster and then move on. O'Connor (1991) says that a typical poster reader will stop, read, and move on—all in 90 seconds or less. You are competing, then, with other activities at the meetings, with tired feet and eyes, and with viewer's time. However, the audience wouldn't be there if they were not interested in your poster and your subject. Design and present your poster in such a way that you keep their interest and so that the experience is worth their time and yours.

Woolsey (1989) suggests that your poster audience can be categorized into three groups: (1) colleagues who follow your work closely, (2) those who work in the same area but not on the same specialty, and (3) those whose work has little or no relationship to yours. He suggests that the middle group is your target audience. Those in the first group will probably stop by your poster to see how you have presented information they already are familiar with and to see if you have come up with any new data or details. Viewers from the third group are not the audience that you need to attract anyway; they simply pause to catch your main point and, not being interested, move on. Members of the second group are interested in your subject and are not familiar with your data. If your poster is clear and attractive, they will probably stop longer than O'Connor's (1991) 90 seconds. To accommodate all three groups, your poster should be

- **Brief and clearly organized**
- **Simple with an obvious central point**
- **Easy to read from 1 or 2 m away**
- **Attractive and aesthetically pleasing**

A closer look at the details that go into a poster can help you satisfy these criteria.

TEXT

The poster format demands concise presentation of information and clear coordination of words and visuals. The content should be a full but brief report. In contrast to the written paper, the poster may omit the abstract, provide little discussion, and use more photographs and color. If the abstract is printed in proceedings of the meeting, it is not needed on the poster unless the sponsoring society requests that it be posted. Some viewers may peruse the display when you are not present. The poster should reflect your credibility as an author by providing substantial justification for your objectives and by giving meaning to your results and conclusions, but extensive written discussion is out of place. Visual communication with pictures and other illustrations can be valuable. You should be present at some point to answer questions and discuss issues, but be sure the text of the poster will stand on its own. See the sample text for a poster in Appendix 13.

As you plan the content for your poster, the following ideas can aid in the audience's understanding.

- **Let nothing distract from your scientific message.**
- **Use short expanses of text and short paragraphs (more than 20 continuous lines will tax the audience's patience).**
- **Present lists when possible especially in such sections as the objectives or conclusions.**
- **Use visual imagery and mix it freely with the text.**

• **Select type style and size, colors, and spacing that make reading the text easy and pleasant.**

• **Provide appropriate handouts such as a summary or abstract with your name and address attached.**

With these points in mind, carefully consider the content and organization of your poster. You are not writing a paper for publication. A full paper can be more complex because it can be read from a comfortable sitting position and can be studied over a longer time. Writing the poster content into a concise, meaningful text without making it too long is a chief concern for anyone designing a poster. One suggestion might be to expand the abstract rather than trying to condense the paper. McCown (1981) suggests that the poster be considered "an illustrated abstract of a publication." The concise, precise format of an abstract should be carried into the text of the poster. To avoid long stretches of unbroken text, break in before it goes beyond 20 lines (10 is better) with a new heading or a picture, table, figure, or other illustration. Long-running paragraphs are formidable, and few in your audience will read them through. Concise lists should substitute for running text wherever possible. The author can carry some of the burden of expanding any discussion orally, and handouts of additional materials can be made available.

No matter how concise you make your poster, most of your audience will not read it from the beginning to the end. They may read conclusions or objectives first, or they may start by viewing data in tables and figures. Some may begin with the introduction, but especially if the text is long and hard to read, they will begin to skip around among parts. For this reason, the poster should be carefully organized, and each section should carry your central point of emphasis.

Organization for the poster should follow the same Introduction, Methods, Results And Discussion (IMRAD) convention that is used for other scientific writing. Let the **Introduction** justify your study and present the objectives. Be sure your hypothesis or purpose is stated immediately and that the **Objectives** stand out in a list to themselves with a headnote or bolder print to call attention to them. Keep the **Methods** section brief unless your purpose is to present a new method. Wherever you can, use a picture, a flowchart, or a list to describe equipment or steps in a process. Just be sure that the methods section provides enough information to make your study credible and to make clear how the data were derived.

Once the objectives are understood, **Results** are the most important part of most presentations. Limit the text, but use clear tables and illustrations for the data. Present only enough data to make the point in your results. Use text as necessary to explain or draw attention to a significant point, but be sure tables and figures are clear without the text. **Discussion** should be limited even more than the methods and results and is often included with the results under a single heading of **Results and Discussion.**

Emphasize the **Conclusions** under a separate heading and keep them brief and in a list if possible. If you cite literature in the text, be sure to provide **References;** these can be listed in smaller type and in a less prominent position than other sections of the paper.

TYPE SIZE AND STYLE

Legibility is of primary importance. Just as the message on a slide must be legible from the farthest seat in the room, so must the message on the poster be easy to read from the customary standing position of the viewer. The title needs to be legible from a distance of 5 to 10 m and the text from 1.5 to 2 m. Blocky, thick styles of letters no less than 23 mm for titles and 5 mm for text will accommodate this requirement for legibility (Table 17-1). Common fonts that

TABLE 17-1
Recommended Type Sizes for Posters[a]

Sample[b]	Font sizes	
	Shown (height in mm)	Range
Title	120 (30)	90–144
Heading	60 (15)	30–90
Subheading	30 (8)	30–60
Text	24 (6)	16–30

[a]This table was originally published by Davis *et al.* (*J. Nat. Resour. Life Sci. Educ.* **21**;158).
[b]All examples are in Univers style.

are good for titles include Swiss, Helvetica, Arial, Univers, or Triumvirate. For the text any one of these is good, or you may choose a conservative serif type such as Times Roman, Bookman, Palatino, or similar fonts. Avoid ornate or script styles, and use italics only where grammatically or scientifically required. A mix of capital and lowercase letters is easier to read than all capitals. As O'Connor (1991) and others suggest, you should use capital letters only where they would be used in conventional texts. The principle of giving the audience what they expect holds true here; our eyes are accustomed to lower case with portions of the letters extended below lines and to different heights above the lines.

COLOR AND PHYSICAL QUALITY

Color, depth, quality, and texture of materials can contribute to the physical appearance and the communication of the poster, or such features can serve as distractions from the scientific message. Refer to Imhof (1982) for principles of communicating with color (Appendix 10). Subdued or dark colors for matting are better than brilliant, intense ones. They can be pleasant without calling attention to themselves and away from the text. Subdued blues, browns, earth tones, greens, or grays are good for the framing. Background paper for the text and the illustrations should be a very light color or white. An off-white, beige, parchment, or other similar light color is comfortable for the eyes.

Other color can be added by highlighting with a bright red, yellow, blue, or green. You may wish to highlight a column in a table or a line in a graph, or even a sentence or phrase in the text. You may wish to use bright colors for bars in bar charts, lines in graphs, or other illustrations, but be discriminating.

Too many colors on one poster are distracting, but color can be used as a point of unity to code portions of the paper. Two different points, objectives, or experiments may carry two different but related colors (for example, blue–gray and blue–green) throughout the poster to code them as separate. Double mounting with contrasting colors can provide trim, can highlight colors in a photograph, and can draw attention to a main point of emphasis. Color is best used when a purpose beyond that of simply attracting attention to itself is evident.

Poster board or matting materials are more durable and more attractive than thinner construction or poster paper. Many matting materials have a textured quality that is attractive. Some are far more expensive than others. Good judgment relative to color choice, sizes of individual pieces, spacing, and placing of materials is more important than the cost of the matting material, but high quality can increase the durability of the poster and the ease of construction. A third dimension can be added by mounting some text or illustrations on thin foam board or cutting the edges of thick matting board at a 45° angle. To make your poster pieces more durable, you may wish to have them laminated, but be careful. A flat finish can blur your text, and a bright laminated finish can create an uncomfortable glare to the eyes in a brightly lit exhibition hall.

SPACING

Before you attempt to make any layout for your poster, find out the exact dimensions of the display board that you will be using. Then construct and place the poster pieces so that they do not crowd the edges of the board. Communication can be enhanced by the sizes and shapes of poster pieces as well as their positioning on the board. Too many small pieces can give a "busy" appearance to the composition, but one large block with all the blank space at the periphery can be equally unattractive. Blank space is important. Woolsey (1989) has said that ideally 50% of the poster should be blank. This blank space can be used effectively to separate parts of the poster and to communicate relationships among the parts.

A good idea is to section the poster into modules based on the organization of the material and space the sections in no more than four or five blocks on the board (Fig. 17-2). Logically, a section spaced at a greater distance from the neighboring section is less closely related than one placed nearer. The physical appearance of the poster should give a sense of unity to the work. Woolsey (1989) also suggests that "the eye looks for edges." Too many edges or small pieces mounted separately tire the eyes. To avoid extra edges, construct the poster in modules or strips and place subheadings or captions for figures and tables in the same frame with the accompanying text or illustration. Shapes that are not expected (triangles, jagged edges, or cutouts) attract attention to themselves. As with highlighting, unusual shapes used effectively may contribute to the communication, but overuse or inappropriate use will simply distract. Shape and arrangement of parts as well as size and color can serve to draw attention to points of greatest importance, to subordinate secondary material, and to unify the entire poster.

With use of conventional headings (e.g., Objectives, Methods, and Conclusions) and with parts of the poster grouped logically and matted with the same color in a limited number of pieces, the viewer should be able to follow the basic principle of Western languages by reading from left to right first and then from up to down. In some instances the presenter may need to code the progression of the paper with numbers or other symbols, but these symbols will be superfluous for posters that are clearly unified with spacing and designed with conventional headnotes and a logical flow of information.

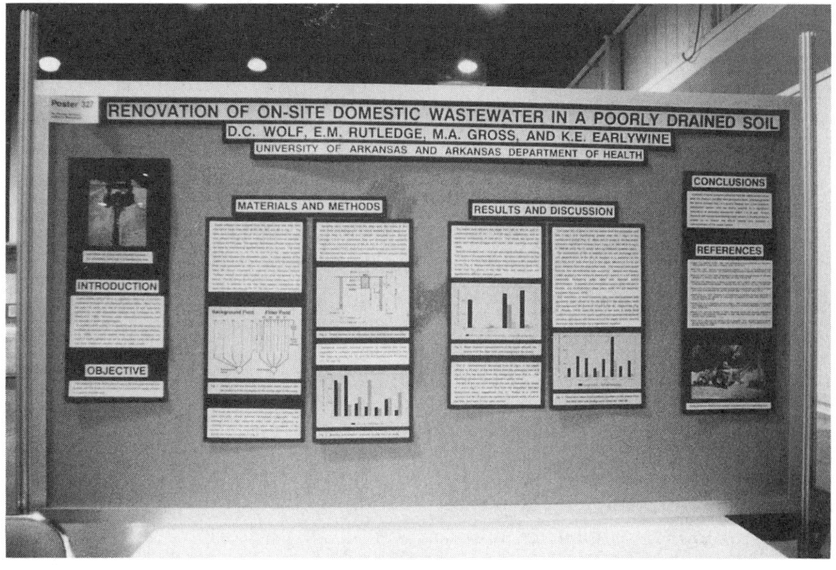

FIGURE 17-2
Divide the poster into a limited number of sections.

PRESENTATION OF DATA

Points of unity and the relative importance of parts of the paper also apply to the presentation of data whether it is in tabular or graphic form. Size and spacing in a table with a limited number of data points can effectively convey a scientific message. Ideally, a table would contain no more than 20 items in the field. Highlighting with color can draw attention to an important column in a table, but as with other attention-getting devices, overuse negates the value.

The same principles are true for graphs. To emphasize a point, the number of lines or bars in a graph must be limited. No more than three lines or six bars are best. The main point that the data carry should be illustrated as clearly and as simply as possible, and any additional supporting data or subpoints should be reserved for the oral discussion or for a more extensive journal manuscript. Color and size in graphs are very important. Too many colors and sizes are confusing, but a highlight of color or a line that is thicker than the others can make a point more immediate. As in other communications, it is important to maintain consistency in labeling from one graph to another. If a point is represented by a color or shape in one graph, the same color or shape should be used to represent that point in other graphs on the same poster.

Photographs will help to make the poster attractive and serve as relief at breaking points in the text, but they can carry an additional purpose in the communication. An appropriate photograph can help the title to convey the subject of the poster or can illustrate visually what the data mean. Photographs, like other illustrations, need to be clear and large enough (at least 5×8 or 6×10 in.) to be immediately comprehensible and make a point relative to the scientific message. Too many or undersized photographs contribute too many edges for the eyes. A matte finish is probably better than a glossy finish for most exhibition rooms in which lighting is usually bright. A well-chosen color for matting or double matting a photograph can highlight the important subject matter in the picture and be aesthetically pleasing.

THE PRESENTER

The relative informality of the poster situation should not relieve a scientist of the responsibility for clear communication and a professional attitude. Your knowledge of the subject, your candor in discussing the science with others, and your appearance and attitude are important to the presentation. The professional meeting will offer many distractions for you as well as your audience. It's your responsibility to be with your poster whenever you are scheduled to be; some of your audience will make a point of being there to talk with you. You may also face the distraction of friends and acquaintances stopping by to chat about things other than the poster. Don't neglect your poster audience for the social audience. Quickly make arrangements to have the friendly chat later, and go back to the poster audience.

There may be periods when no one approaches your poster, and you feel that you may as well leave. Not so. Audiences are likely to appear one or two at a time. Don't be discouraged if masses of people don't flock to your poster. One important attraction of the poster technique is that the audience is limited to only the truly interested. If your poster adheres to the criteria for a good poster and if a few people read most of it and talk with you, you have been successful. Whether the audience is one or a dozen, execute your professional role with clear communication, knowledge, and sincerity.

HANDOUTS

A simple handout can enhance the poster communication. The authors' names, addresses, and phone numbers along with an abstract or other condensed facsimile of the poster, a list of pertinent references, an important method, or a table or figure may prove valuable to the viewers after they leave the meeting.

MAKING IT FIT

Poster display boards differ in size at different meetings. Some are as large as or larger than 4 × 8 ft (ca. 1.3 × 2.6 m) and some as small as 3 × 3 ft or 1 × 1 m. A 4 × 6-ft (ca. 1.3 × 2.0-m) board is also rather common. Some are mounted on a stand or rest on a table and are oriented with the longest dimension horizontal. Others may have the longest dimension vertical and stand on the floor. Before you begin to construct the poster, you must know the space and orientation you will have for your display. Decide how much material you can display effectively in the space allowed. Keep as close to eye level as possible, and if your display board rests on the floor, it is wise not to use the lower portion that would be difficult to read.

In addition to space limitation, you must recognize that, with dozens of other posters also on display, yours will command a limited amount of viewers' time before they move on to others. The most common communication fallacy that I have observed in posters occurs in the attempt to present too much material in the time and space allotted. This problem manifests itself in both too much text and too many data. Be willing to make one or two points and leave your other information for future papers.

In making the information fit the board, the second most likely fallacy is in poor construction that distracts from the scientific message. Keep the display simple and attractive. With colors, shapes, and illustration or with words themselves, no device should be used that is merely decorative and attracts attention to itself and away from the science. Of course, you want your poster to be aesthetically pleasing, but foremost you want the scientific message to be clear. These two objectives can complement each other. As Imhof (1982) says, "Clarity and beauty are closely related concepts." Poor or inappropriate use of color is especially noticeable in posters. Too many, too brilliant, too pale, or uncoordinated colors distract from the scientific communication in a poster as sure as a clown suit or gym shorts would distract from the words of a speaker at a scientific meeting.

Standards of communication for any single visual, such as a chart, a slide, or a photograph, are also the standards for the entire poster composition. The completed poster must be able to stand on its own with all communication devices in place. The materials should be fitting for the poster format, and the amount of content and the number of illustrations should not overload the message. The display should command credibility, communicate only a limited number of points, and be accurate, clear, legible, concise, and aesthetically pleasing.

In Fig. 17-2, notice the limited text (Appendix 13). Much of the information for methods and the data for results are presented in illustrations and figures. The objectives and conclusions are set apart to be readily visible, and a prominent title and the headings can be read from a reasonable distance. The entire poster is arranged in modules with spacing helping to carry the communication. For its format, this poster deserves a high rating as a scientific presentation.

CONSTRUCTING THE POSTER

Rapidly changing computer technology makes it impossible to predict what you or I may be using to construct our next poster text. Large lettering, graphically plotted data, and illustrations can be produced by computer. Fact sheets and computer company representatives can inform you about what programs are available, functions they can perform, and which pieces of equipment are compatible.

Once the text is composed and photographs and illustrations are ready, final construction can take place. If you happen to be lucky and wealthy, you may have a graphic artist construct the poster. More likely, you'll put it together yourself, and with a few instructions you can do a good job.

Before you begin, have your final text copied onto heavy (65-lb, index weight) paper. This paper is durable, it takes gluing without wrinkling, and the color of the mat board will not show through to blur the type. Cut the mat board large enough to provide a frame around each section of the text. Then glue the text onto the mat with a quality adhesive. Liquid glues, spray adhesives, two-sided tapes, or even carpet tape can be used. You may want to test the ability of the adhesive to hold and endure before putting the entire poster together.

Remember that you have to get the poster to a meeting, and you will need to pack it in a relatively small case to carry it onto a plane or even to transport it conveniently from a hotel room to the poster session. You can mount the text on small sections of mat board and simply join the edges when you attach the pieces to the poster board, or you can construct columns or sections of material, cut them into smaller segments, and use duct tape or strapping tape to hinge the back side so that you can fold them for packing and simply unfold them to place the connected pieces on the board. This technique is especially good to use with titles that stretch across the width of the board. The title section and each column in Fig. 17-2 were hinged in this way.

You will acquire other tips on construction by trial and error. I've picked up an assortment of ideas along the way that may be helpful to you. For example, don't spray too much adhesive in a closed room; you may develop a headache or more serious reactions. Use Velcro to attach poster pieces to the board if possible; pushpins are more trouble and less attractive. Be sure your cutting tools are sharp when you begin to cut paper or poster board, and check your cutting surface. You wouldn't want to scratch your advisor's lab benches or a table. Cut against other poster board or cardboard or an artist's cutting board. To keep edges straight, hold a metal straightedge against your marked matting, and for cutting long edges ask someone to help hold the straightedge. Watch the position of all hands and fingers in relation to the knife. Measure carefully before making a cut. Even small variations in frame widths or paper size can look magnified on a poster display. Finally, if possible, talk with someone who has constructed a poster before you begin the task.

TIME

In constructing posters, time is the antagonist. As you conduct your research, keep in mind that you may need to construct a poster in the future. Take photographs for prints and isolate the data that you will use. Weeks before your poster presentation, you can have photographs enlarged, set up tables and figures, and write the text. If you also need to be working at other tasks, allow yourself at least 2 or 3 weeks after the text is ready to put the poster together. Allow time to reconstruct after reviews. Don't mount the text on mat board until you have had it reviewed by several colleagues. As with other forms of communications, reviews and revisions are essential.

Several sources of information are helpful in learning to design and display posters. Basic principles in visual communication that are valuable, not only for posters but also for other visual displays, can be found in a book by Reynolds and Simmonds (1983). The best and most recent concise advice I've found on the poster format is that of Woolsey (1989), and O'Connor (1991) also includes some brief but good advice. Other information is available from almost every society that uses the format.

References

Imhof, E. (1982). "Cartographic Relief Presentations." de Gruyter, New York.

Maugh, T. H., II (1974). Poster sessions: A new look at scientific meetings. *Science* **184,** 1361.

McCown, B. H. (1981). Guidelines for the preparation and presentation of posters at scientific meetings. *HortScience* **16,** 146–147.

O'Connor, M. (1991). "Writing Successfully in Science." Harper Collins *Academic,* London.

Reynolds, L., and Simmonds, D. (1983). "Presentation of Data in Science." Nijhoff, The Hague.

Woolsey, J. D. (1989). Combating poster fatigue: How to use visual grammar and analysis to effect better visual communications. *Trends Neurosci.* **12,** 325–332.

18
GROUP COMMUNICATIONS

"The true spirit of conversation consists in building on another man's observation, not overturning it."

BULWER LYTTON

Whether you work for a private industry or a public agency or institution, you will find yourself involved with group communications. As with individual efforts, preparation and competent execution are important to provide clear communication and prevent wasted time and effort. Selecting a format as well as preparing for and carrying out communication with a group depends to a large extent on whether an audience is involved.

GROUP COMMUNICATIONS WITH NO AUDIENCE

You may be asked to serve with closed groups to make plans for research projects, to decide policy, or to evaluate a fellow employee's progress. These professional duties are usually performed without an audience and in the format of a round-table discussion or a board or committee meeting. The **round-table discussion,** or **buzz session,** is informal. The purpose may be for brainstorming or for a professional sharing of ideas with no formal agenda followed and no decisive outcome expected.

A **board** or a **committee** is a more formal, small group selected from a larger organization to assume a given responsibility. Participants follow agendas and expect outcomes that are usually reported to the larger body through their leader. **Standing committees** serve for indefinite periods; **ad hoc committees** are set up to deal with one particular job and are disbanded when the job is complete. Definitions begin to overlap with the use of the term **task force,** which is a kind of ad hoc committee but is usually made up of experts in a given area whose job may require very detailed coverage of an issue.

Whatever format is used, professional ethics dictates that you perform actively to the best of your individual ability to accomplish the group's goals (Wilson and Hanna, 1993). Sometimes individuals are appointed to serve in such roles as chair, recorder, or even a devil's advocate to search out problems. Often a group selects from among themselves the persons who will fill such roles, or a given personality may assume a role of leader or antagonist without being chosen for the position. Once the group is selected, the task with no audience involved is often decision making or problem solving.

Decision making involves alternatives. Defining what the alternatives are may require research and discussion. Making the best possible decision can depend on gathering information and even solving problems along the way. But after discussion of the pros and cons of each alternative, the group should reach a consensus or conduct a vote to make the final decision.

In **problem solving,** there may be no obvious alternatives. The best solutions can be reached with a step by step procedure. As with the scientific method, we begin with a problem, identify or define it clearly, gather data, analyze that information, synthesize the new information with what is already known, and evaluate the outcome. Reaching a solution in a group involves coordinating the expertise and opinions of several minds and personalities. The group has the advantage of dividing work loads and contributing multiple expertise but the disadvantage of coordinating diverse opinions. Recognizing these conditions, the group must follow essentially the same steps as the individual would follow.

Procedure for Group Problem Solving

1. The problem is clearly defined and objectives are set forth and understood by all members of the group.

2. Members of the group plan their individual and collective actions. They may divide responsibilities for gathering information and offering opinions.

3. As individuals and as a group, they devise a plan of action.

4. They act on the plan and analyze outcomes.

5. They evaluate the results of their actions and determine whether the solution was acceptable.

With the group, as with individual problem solving, a poor solution to the problem can result from inadequate or faulty information. Each step in the process must be fully executed in order that the next step can be achieved.

Groups sometimes get in a hurry to reach the goal and fail to define the problem fully or plan their actions carefully before they proceed. Without deliberate planning, members of the group will begin to stumble over their lack of understanding or the sequence of actions, and then they must return to the definition of the problem, explain it, and map plans all over again.

The size of the group and the time given to accomplish a goal are crucial to success. A group of more than 10 or 12 people will have trouble involving all the personalities and all the individual ideas. An odd number of 3, 5, or 7 can usually work well together. Planning contributes to effective use of time, but because of the need for input from several individuals, the group activity will take longer than communication by a single individual. However, the combined expertise of the group can make the additional time worthwhile.

If your group is too large, **brainstorming** is one way to make progress. The advantage of any kind of brainstorming is that, at this beginning, ideas need not be criticized but just presented for consideration. The entire group might brainstorm until they reach a consensus on what ideas to discuss, but if many personalities are involved, the less assertive members of the group may be intimidated. The large group can divide into smaller buzz groups for preliminary discussions and then present their ideas and objectives to the full group. Another brainstorming technique for handling a group that is too large is to ask the entire group to be silent for a few minutes while each participant writes on a card or slip of paper what he or she believes are the most important ideas to consider. Then a recorder or leader will present all these ideas to the entire group. From the list, ideas can be combined and organized and priorities can be established. When specific topics or objectives have been filtered from the random ideas that evolve during brainstorming, then the critical process begins.

Positive progress can be made only with thorough, critical analysis by individuals. As a member of the group, you should feel responsible for researching whatever information is pertinent to the decision making, the problem solving, or the presentation of information. You shirk an ethical responsibility if you come unprepared and expect to know enough already or to let others make the decision. When individuals do not assume individual responsibility for preparation, information, and participation, the group interaction can often degenerate to confusion or, perhaps worse, to a point where individuals are not thinking for themselves (Fig. 18-1).

Janis (1982) coined the term *groupthink* to apply to a mentality of group agreement without the individual minds sorting out the best decision or solution to a problem. "Groupthink occurs when a group strives to minimize conflict and reach a consensus without critically testing, analyzing, and evaluating ideas" (Beebe and Masterson, 1994). Generally, a group should be cohesive and avoid the kinds of conflict that are destructive and disruptive, but healthy conflict and debate are important to decision making or problem solving. Passive group members may believe they are being cooperative and agreeable

FIGURE 18-1

Groupthink involving too many empty minds does not provide the best solution to a problem.

when, actually, conflict or a devil's advocate would be more valuable to decision making and problem solving. With open, critical, professional discussion and debate, the best ideas can be sorted out over time and the appropriate conclusions can be reached.

The group effort is no better than the composite efforts of the individuals in it. In working with a group, be critical and cooperative at the same time. A professional can be critical of ideas without being critical of the people offering those ideas. Cooperation demands that you be open-minded, that you listen as well as talk, and that you give and take opinions; it often requires that you allow your ideas to be criticized, that you criticize the ideas of others, and that you be willing to compromise on points of disagreement. The passive power of groupthink can obliterate the best possible ideas. Unanimity is a good quality. It should be strong once a decision is made, but it is a end, not an objective means. Individual responsibility is important whether or not your group is working with an audience.

GROUP COMMUNICATIONS WITH AN AUDIENCE

An audience adds another dimension to group communication. Members of the audience are active participants even if they only listen. With the audience involved, group communication can take several forms. The **panel discussion,** carried out by a small group in front of an audience, is usually unrehearsed but need not be unorganized. The panel employs informal discussion with a leader to moderate the communication. It may appear to be spontaneous, but that appearance is usually the result of careful planning. A **symposium** is also made up of a small group, but participants act as individuals with each making a prepared speech before the audience. Their speeches are on the same or closely related subjects. The symposium is more formal than the panel discussion with little or no discussion among the participants. A symposium often turns into a panel discussion among the speakers or a forum with the audience. A **forum** involves all the attendants at a meeting, i.e., total audience involvement, under the direction of a leader. The town meeting is a forum.

When an audience is involved, the planning for group communication is extremely important. Every group participant must clearly understand his or her role, how much time is allotted to the entire program, and how much each participant should use. It is equally important that members of the audience understand their role. Can they ask questions or make comments during the discussion, after the discussion, or not at all? A good moderator or chair will explain to the audience and the participants the goals, procedural plans, and individual and group roles. Determine your individual role as leader or as a member of the group, and take charge of your responsibilities. Let me illustrate these responsibilities relative to the panel discussion.

The Discussion Leader

If you are in charge of the panel as chair or moderator, before the meeting you will need to carry out several duties:

1. You may be responsible for choosing the panel members. Choose carefully. Participants should have expertise on the topic to be discussed; they should be articulate and capable of presenting organized information. If a subject is controversial, they need to be able to maintain a professional attitude that does not allow for emotional outbursts or argumentative assaults on other group members. In other words, when you select participants, carefully evaluate their expertise, personalities, and communication skills. Although you may rightfully have your own bias about a subject, balance the membership of the panel so that all sides of a question can be given equal consideration. Two conflicting personalities with two conflicting opinions can create a problem if both parties cannot conduct the discussion in an open, calm, and professional manner, especially if the discussion leader demonstrates a bias.

2. When you have chosen members of the discussion group or have accepted leadership of a group already chosen, you need to be sure you understand the purpose and the objectives to be put forth and accomplished. You may wish to consult with each participant for opinions to incorporate into the

program. Study these opinions and your own carefully in view of the objectives and plan a strategy for conducting the panel discussion.

3. Once your members are chosen and your plans are clear to you, be sure that each of the participants understands the objectives, the procedural plans, and the role he or she is expected to play. Talk with each personally if possible, and send each a written agenda or outline of the program with names and roles of other participants and with suggestions for what you expect.

4. Consider the external influences on the communication effort. These include the audience, the room, and other physical conditions; the time involved and how portions of time should be allotted; and any other condition that could influence the success of your meeting. Sometimes bad weather, the noise next door, the time of day, or lighting can destroy an otherwise successful plan. Even the arrangement of the chairs in which participants sit will influence the meeting. A curve-shaped arrangement is often better than a straight line. The room can be too large or too small, microphones may be needed, or your program may overlap in time with other sessions of interest to the audience. Your audience may be sleepy or belligerent. To whatever extent these things are under your control, control them.

At the meeting it is usually your responsibility as leader to take charge of the situation. Arrive a little early and take care of final details in the physical setting and greet the participants as they arrive. Your own confidence and relaxed attitude can be contagious. Carrying out the communication smoothly will also depend on how well you follow the organized plan. As leader, consider the following responsibilities.

1. Introduce the session to the group and to the audience if one is involved. You will introduce the panel members to each other, if they are not already acquainted, and to the audience. Then, you will outline the objectives for the discussion to both the panel and the audience. At this point be sure that members of the audience understand the role they are to play. Let them know whether you invite interruptions or prefer that questions and comments be saved until panelists have completed their initial comments. If you fear that the audience will be too vocal, you may wish to ask them to remain quiet throughout or to submit any questions they have in writing (perhaps on cards that you provide). A smooth beginning can establish a rapport with the audience and contribute to a successful program. Despite your careful planning, if an audience is involved, be ready for surprises.

2. Your duties as leader during the session are to participate, moderate, arbitrate, evoke discussion, keep discussion on track, and summarize at appropriate junctures. Encourage all panelists to express their views; prompt the more timid with questions, and pull the conversation away from the aggressive talker. If an individual tries to dominate or if the subject gets off track, inter-

rupt the discussion, summarize the points that have been discussed, and then redirect the group toward a question or a new point. If the discussion becomes too nebulous or general, tactfully call for details to support statements being made. In a relatively long session, it is good, even if the group is staying on track, to inject a summarizing statement more than once. This kind of control may be important in order to keep the program within a designated time as well as to be sure all issues in the original plan are discussed.

3. You will also be responsible for concluding the session. Again, summarize the points discussed and any conclusions that have been reached. Often you will need to reiterate the objectives and show the extent to which they have been fulfilled. It will be important to include in your summary those unexpected points that arose during the session. In other words, you cannot completely plan your concluding remarks before the session begins; you must take notes throughout the discussion and quickly organize and present major points that were brought out. The job of concluding will be easier if you carried out your other responsibilities effectively.

Finally, in ending the session, you will include an expression of appreciation to the panelists and to the audience for their attention and participation. If further sessions are to be held or announcements made, you may be in charge of information about those. Always conclude with a professional and pleasant attitude as leader, even if the discussion itself has deviated from those qualities.

Responsibilities of Group Members

Leadership skills are essential for every panelist. Members must consider the audience, the individuals involved in the discussion, the objectives and plans for the meeting, the time involved and how much of it belongs to each of them, and the professional attitude essential for a successful meeting. As a member you may be expected to take the lead on certain issues or you may be in charge of a segment of time. Often, the designated leader or moderator may be put into an uncomfortable position by a question from the audience or a point being discussed about which he or she has previously developed a reputation. At such a point, a panel member can supply the needed support to keep the discussion on track by injecting an apt comment or temporarily and unobtrusively assuming psychological leadership.

All members, including the leader, must work as a group and as individuals. Each and all must be responsible (1) for defining, interpreting, and analyzing the subject and points that arise; (2) for keeping the discussion relevant and alive; (3) for evoking discussion from each other and, if applicable, from the audience; (4) for controlling unproductive conflicts and giving each group member equal opportunity to participate; (5) for giving each issue appropriate time and attention; and (6) for following plans set forth in the preparation for the meeting and helping to summarize the issues at various points along the way.

PLANNING FOR DISCUSSION GROUPS

Organization for group communication includes the **subject matter,** the **time,** and the **people** involved. The following are possible structures on which you might base your plans for a group discussion. These examples apply where both a small group and an audience are involved, and they are certainly not inclusive of all possibilities.

The Forum

In a short statement, the chair may introduce the subject to be discussed and outline the goals to both the group (who should know already) and the audience. This leader then opens the discussion to both the panel and the audience. This technique turns into a forum or becomes a question-answer session with the panelists serving simply as leaders along with the moderator. The method is valuable for public meetings in which experts are invited to serve as a panel of consultants. The danger in this structure is the difficulty in controlling a large group, especially if an issue is emotional or controversial and the audience includes some aggressive, opinionated personalities. All panel members need to help control the situation and to move toward a prescribed goal in a professional manner. The leader should be a strong individual who does not allow the meeting to get out of control. Control in such situations comes more with finesse than with demands.

The Closed Panel

The chair may introduce the general subject in a short speech and then turn discussion over to the panel with no audience involvement. This technique is often used with television or radio broadcasts. The chair keeps the issues moving by asking questions and by directing and redirecting the discussion from panelist to panelist in an equitable manner. If a live audience is present, they may insist on being involved. Panelists should be careful not to encourage audience involvement. Again, a strong leader must assert his or her position, outline the plan to the audience from the beginning, and divert any interruption that could keep the meeting from reaching a goal.

Symposium/Panel/Forum

The chair may introduce the general subject and the basic issues to be discussed. Then, each group participant may also make a short speech in which he or she outlines one of the issues or one point of view. To this point, the format is that of a symposium, but it may develop into a panel discussion or a forum after the short speeches.

Decisions on which direction the meeting will take should, tentatively at least, be made before the meeting. Leadership roles may be divided among panelists by subject matter or points of view, and each of these leaders must assume responsibility for moving toward a common goal. Audience involvement may require that the leader serve as moderator between the panel and the audience. The symposium that develops into a panel and forum is valuable for controversial issues or for subjects that should be considered from several points of view. Success depends on a controlled, professional atmosphere.

Whether or not an audience is involved, prepare and organize materials for group communications as carefully as for individual presentations. A great deal of time can be wasted if discussion leaders are not prepared with an agenda or outline for conducting a meeting and if participants have not consulted that agenda and organized their thoughts and supporting materials to express their views. Keep in mind several important points that can make group communication successful.

- Set a specific goal but keep plans simple. With group interaction you need to limit what you can accomplish in a reasonable time.
- Start on time, give each issue appropriate time, and end on time.
- Be sure everyone, including any audience, knows what format is being followed and what goal is pursued.
- Think as an individual; offer critical analysis and clear evaluation of information discussed; and avoid groupthink.
- Work toward the prescribed goal, summarize along the way, and avoid digressions.
- Maintain a professional attitude and cooperate with the leader and other participants.
- Sustain equitable participation; don't talk too much or too little.
- To the extent that you can control such things, be sure the physical situation is comfortable for everyone and conducive to good communication.

The group structures I have discussed clearly do not exhaust the creative possibilities, and your choice of format for communication may depend on the subject, how formal or informal you need to be, how much time you have, and at what point you wish to involve an audience.

Recent technology has provided us with network discussions and conference calls that may or may not allow us face to face exchange with other members of the group. Group interviews may take place on television or radio with an audience receiving the broadcast but not in the studio, or a televised panel can appear before a live audience. Whatever the situation, perhaps the most common group communication for the scientist is with a team of fellow researchers.

Whether it is a committee appointed to recommend a company policy or a panel of experts discussing an issue before an audience, the principles that take group communication to a desired goal are the same. Your participation should be active and ethical. You can work professionally with personalities in your group if you assume the proper role for your own personality and acknowledge the rights of others. Think independently and respect the independent thinking of others, even those who disagree with you. If you have trouble working with a group, study information on group communication such as that found in Beebe and Masterson (1994), Wilson and Hanna (1993), or Jensen and Chilberg (1991).

References

Beebe, S. A., and Masterson, J. T. (1994). "Communicating in Small Groups: Principles and Practices," 4th ed. Harper Collins College, New York.

Janis, I. L. (1982). "Groupthink: Psychological Studies of Policy Decisions and Fiascoes," 2nd ed. Houghton Mifflin, Boston.

Jensen, A. D., and Chilberg, J. C. (1991). "Small Group Communication." Wadsworth, Belmont, CA.

Wilson, G. L., and Hanna, M. S. (1993). "Groups in Context: Leadership and Participation in Small Groups," 3rd ed. McGraw–Hill, New York.

19
COMMUNICATING WITH OTHER AUDIENCES

"Rather than regard such activities as irrelevant distractions, we must realize that informing the public is a vital aspect of our jobs."

RICHARD S. NICHOLSON

In your scientific career, one of your most serious responsibilities is to communicate with those who know little about your area of expertise. The medical doctor and the science teacher face this responsibility every day. We sometimes complain about scientific illiteracy among the public. It may be that those who are scientifically illiterate have not taken advantage of educational opportunities, or it could be that scientists do not provide them with adequate opportunity to acquire such literacy. Keeping scientific knowledge enveloped in an aura of mystery is not healthy. As any good scientist can attest, the unknown elicits curiosity, but it can also evoke emotions of anxiety and fear. Help the nonscientist acquire enough information to satisfy some of the curiosity and to allay the fear.

In the first place, be proud of what you are and what you do. Progress in science is remarkable, and you want your audience to care about your science. To reach them, try to broaden your own mind and see values from their vantage point. Scientific and human values often differ and sometimes even conflict. Science is not the only possible constructive truth in the world, and every scientific experiment is based on limited measures. "Any single set of factors can be interpreted in a variety of ways" (Gould, 1981). Science has produced both the good and the bad. Your belief in science can be firm, but someone else may believe as strongly that science is a wayward and destructive body of knowledge. Neither of you knows the full truth. Scientists are not amoral, and the products from science are not immune to other human values. Communication is the avenue through which you can reconcile differences so that those with differing opinions can work together for some specific good.

It's easy to ignore opportunities for communicating with nonscientists. You don't have time for auxiliary tasks that will probably provide little ammunition for your promotion. Besides, you have seen how "the public" can take a professional's words out of context and make legal lies out of accurate remarks by putting them in a new context. Certainly, scientific, and especially environmental, issues are not detached from politics and everyday life. Public policy on scientific issues often is shaped by the scientific illiterate. In a democratic society anyone who knows about a subject should contribute to the quality and education of a self-governing citizenry.

You can fulfill your responsibility for communicating with audiences other than your peers in science if you will think in terms of who makes up the audience, what avenues you have for reaching various audiences, what they want to know, and what techniques can best convey your messages to those people.

AUDIENCES

Audience makes the difference in where, what, and how you communicate. How do you talk with or write for an audience that has no idea what deoxyribonucleic acid, trinitrotoluene, anthesis, or denitrification is? Above all, don't strut like a peacock simply because you know some words unfamiliar to your audience (Fig. 19-1). You actually speak the same language they speak; you simply have acquired some alternative vocabulary words that make it more convenient to communicate with your peers in science. You should be capable of using a term like "flowering" or "bloom" rather than anthesis or of referring to "TNT, a highly explosive material" rather than trinitrotoluene. If you can't use such synonyms, perhaps **you** have a literacy problem.

I knew an advisor who told his graduate student just before a speech to "snow 'em with the terminology." He contended that, because he and the student were working in a new area of research, the student should be able to string together a vocabulary that even the scientists would have trouble following and, thereby, "leave them impressed." I felt at the time that that advisor

"Excellent communication skills. Poor choice of words."

FIGURE 19-1

Be careful with your choice of words. (Cartoon from Martha Campbell, (1991), *Phi Delta Kappan* **73,** 130. Used with permission of the author.)

should be beaten about the head and ears. In any communication effort, the objective is to make yourself understood, not to "snow 'em."

For a basic understanding of your audience, you need to consider such things as their ages and educational level. They may have never had an opportunity to attend college, and if they are Ph.D.'s in the humanities, they may never have taken science beyond a general biology course. Most of them, however, are as intelligent as you are. You need to think about their interests, their prejudices, their attitudes, and maybe their occupations. Paint a picture in your mind of the typical person in your audience. That person could have high regard for your scientific education and respect everything you say. Let that trust humble you a little. On the other hand, if the typical member of your audience has prejudices or a negative attitude about the values in science, he or she has reasons for such notions whether or not objectivity is at the root of those reasons.

Your job will be easy if the audience is a homogeneous group such as science club students from grades six to eight or the chamber of commerce of a small town. More difficult is "the public" or even a college audience composed of graduate and undergraduate students from all areas of science. In your scientific paper written for your peers, you may say, or assume they know, that denitrification is the reduction of nitrates or nitrites to nitric or nitrous oxide by denitrifying bacteria under aerobic conditions with an available carbon source. You would be able to tell the college students the same thing, but undergraduates might need some further explanation.

For the public, your objective will be different. They probably do not need a scientific definition; they need to associate the scientific idea with their own world. You may then explain that nitrogen fertilizers are essential to growing the food we eat, but too much nitrogen in the soil in the form of nitrates can get to the groundwater, which is the source of much of our drinking water. Too much nitrate in our water supply can be harmful to the health. To break down those nitrates, we need a soil that contains helpful microorganisms that convert the nitrates into less harmful gasses. With this information you are as close to the audience as the water they drink. Most intelligent adults will understand what you are saying whether or not they have sat through chemistry and microbiology classes.

For the kindergartners, you might need to explain even further; the microorganisms may become tiny bugs that we can't even see. Whatever adjustment you have to make in your terminology, there is no reason to suppose that you cannot communicate with a group of people simply because you are more educated in one area than they are.

AVENUES

Identifying the audience and empathizing with them are keys to success with any communication, but how do you acquire such an audience? Opportunities will arise, but you can also open doors yourself. You could join civic groups that would be pleased to have you provide scientific expertise on public problems such as safe water supplies or waste disposal. You may belong to an interest group such as amateur bird watchers who would like you to tell them more about birds, ecology, or other areas that you know better than they do.

Your own company or agency could provide avenues for communication. Your position can require that you write for or speak with lay audiences. A major responsibility of scientists in the Agricultural Research Service and the Cooperative Extension Service is to transfer information about what is going on in scientific research to the agricultural producers and industries. Your efforts may be to inform nonscientists within as well as outside your company. You can make speeches or produce copy for fact sheets, leaflets, technical bulletins, newsletters, policy statements, employee or public pamphlets or brochures, exhibits or posters, and other creative forms. Your employer may encourage you to visit civic clubs or schools or write for the popular press. Calvert (1990) provides helpful information on media and methods.

The media come in numerous other forms. Often, scientists are invited to participate in interviews or other programs on radio or television. Live interviews are a bit scary at first, but good preparation can alleviate the fear. Suggestions by Gastel (1983) can help. Rather than the live interview, you might be more comfortable with other avenues for communication. You might produce your own videotape that can be edited for a given audience such as high school students, first graders, or the public. You can be invited to give a slide presentation or just make a speech to a group of children or a civic club. Electronic communication can now reach many audiences. Children of the future will use a computer network to acquire information that we used to obtain from the encyclopedia. You could contribute to a store of information on that network.

The written media contain even more variety than the spoken. Write for a newspaper or popular magazine. You can submit a feature article or maybe contribute a regular column. Some scientists even pursue careers as "science writers" for the popular press. If you are interested in such a career, read Gastel (1983). But you can maintain your research, teaching, or service position as a scientist and still use the popular press and other media for communication with nonscientists. Both in speaking and in writing, you will find great opportunities for alleviating some of the scientific illiteracy that prevails.

SUBJECT

The subject you select or are asked to write or speak about will depend on the audience. As a scientist, it is in your own best interest to provide information that will help nonscientists understand, trust, and appreciate science. They have a right to know how seedless watermelons have been developed, what chemicals are in their food, or why a hormone can make them feel better and even change their attitudes. They also have a right to know whether genetically altered species might destroy Mother Nature's own creations, why we should bother to protect a blind lizard or a night bird, how to effectively dispose of waste that humans and animals create, whether nuclear energy is as safe as electricity, what side effects drugs may produce, and hundreds of other questions that affect their daily lives, their jobs, and their government.

Again, try to select a subject based on the interests of the audience. It should be timely and important. Clearly identify your objective or purpose in communicating your subject. You may tell a group what they can do to restore their land to productivity after a lengthy flood or help them understand how to guard against communicable diseases. You could encourage them to read and follow pesticide labels before scattering herbicide in the yard or insecticide in the house. Just like you, they are interested in safety, current happenings, unusual events or discoveries, and new developments in science and technology. They are interested in nature and in using and protecting the environment. They care about what science does to help or hurt the economy and their employment.

How you treat the subject can be as important as the subject itself. Be cautious with disclosures about new discoveries. It's better to tell the audience you don't know than to lead them through your own enthusiasm to making faulty presumptions based on what is still experimental. Often when you do not wish to commit yourself to a possible outcome of a scientific discovery, you can refer the audience to original sources of information and let them draw their own conclusions just as you do. Remain positive and congenial if they question your credibility. Don't assume that the audience is unintelligent or has a bad attitude. Principles for good communication require that you respect your audience and take criticism, perhaps from people that know far less than you do about your subjects. React to that criticism in a calm, rational manner. It can help to have someone review what you will say or write for a given audience and then make revisions before you deliver the information.

TECHNIQUES

With a clear purpose and your audience in mind, your task is to relate the unfamiliar to the familiar. You have to start with what the audience knows already whether it be high school science students or preschoolers at your daughter's day care center. You will need to use more extensive illustration than you would with your peers in science. Slides, demonstrations, and other displays can help them understand your points more quickly than words alone will. Don't hesitate to use scientific data for support of what you say or write. Just be sure to present those data in a form that your audience will understand and one that is credible and not fragmentary. Use definitions and comparisons and visual imagery. Compare ideas in the science with something familiar to the audience. For example, if we could see a DNA molecule, it might look like a couple of chains looped around each other. Third graders could visualize that image.

Writing or speaking about science should follow the principles of simplicity as laid down by such authors as Zinsser (1988a). Be sure that what you say or write is accurate and as complete as possible. Be conversational and direct, and repeat your main points for emphasis. Use first and second person and the active voice. Also, watch your tone as well as your vocabulary. Watch out for any note of prejudice or condescension in your diction. It is easy to inadvertently say something that can offend some individual or group in your audience. Avoid complexity and incomplete explanations. Be sure you carry the audience forward with transitions that they understand. These transitions sometimes require more lengthy explanation of what connects two points than would be required with a scientific audience. Narrow your subject to one that can be explained in terminology the audience will understand in the speaking time or writing space you have available.

Besides conveying the unfamiliar in familiar terms with accuracy, completeness, and simplicity, another principle to consider is the interests of the audience. What they care about knowing is as important as what you think they should know. They may not care what a DNA molecule looks like, but they want to know how you can use knowledge of DNA to change characteristics in a tomato or provide evidence in a rape trial. Join them in their world; don't expect them to come to yours immediately. As a good communicator, you can guide them to at least one view of your world if you lead them from their own interests.

Several sources in addition to Zinsser (1988a) can be helpful for exploring the principles of good communication. I've already mentioned Gastel (1983) relative to science writing and Calvert (1990) for information on media and techniques. But you also may need to return to basic instruction in a text such as that of Burnett (1994) for writing or Smith (1984) for speaking. Read some of the popular magazines or newsletters in your discipline; you can find many of them on the library shelves next to your professional journals. Zinsser's (1988b) *Writing to Learn* contains good examples of science writing. Science writers such as Carl Sagan or Stephen J. Gould can teach you much by example. Find out where the kids in your neighborhood are getting information about science. Then make a commitment to communicate science outside the scientific community.

You don't have to be a Carl Sagan or Jacques Cousteau or Stephen J. Gould. You cannot provide all the answers to all the public. But if each practicing scientist would contribute a bit of his or her knowledge in this regard, scientific illiteracy would not be such as problem. In fact, it could well be that we would all take better care of that world of physical matter that is the subject and breeding ground for science itself.

References

Burnett, R. E. (1994). "Technical Communication," 3rd ed. Wadsworth, Belmont, CA.

Calvert, P. (Ed.) (1990). "The Communicator's Handbook: Techniques and Technology." Agricultural Communicators in Education. Maupin House, Gainesville, FL.

Gastel, B. (1983). "Presenting Science to the Public." ISI Press, Philadelphia.
Gould, S. J. (1981). "The Mismeasure of Man." Norton, New York.
Smith, T. C. (1984). "Making Successful Presentations: A Self-Teaching Guide." Wiley, New York.
Zinsser, W. (1988a). "On Writing Well," 3rd ed. Harper & Row, New York.
Zinsser, W. (1988b). "Writing to Learn." Harper & Row, New York.

Appendix 1
WEAKNESSES IN SCIENTIFIC WRITING

The most serious problem with most scientific papers is an unwillingness or inability of authors to revise their own work. Failure to polish a draft is a weakness in itself and allows other weaknesses to be too prevalent in scientific writing. If you feel uncomfortable with detecting weaknesses in your own paper, look for some of the following.

LACK OF PREPARATION

Some people begin to write before they are ready or before they have carefully studied a subject. You must have an audience in mind and have something to say to them. Be confident in your own knowledge and opinions, but to achieve credibility, you must know your subject and believe in what you are saying. You need to support that belief with scientific principles as well as information from your own data and other research findings. In other words, prepare by studying and understanding the literature, the science, and your data.

WEAK ORGANIZATION

A paper must be centered around one point of emphasis. That point constitutes the purpose of the paper and gives birth to the objectives or supporting points. A logical pattern and progression in the arrangement of ideas is essential to scientific communication. Almost any scientific report can use the following generic outline as model, but you must organize details within this pattern.

I. *Introduction*
 A. *The Subject and Organizational Points*
 B. *Background and Justification*
 C. *Objectives of the Study*
II. *Materials and Methods*
 A. *Materials and Location of the Experiment*
 B. *Methods of Sampling*
 C. *Methods of Analysis*
 D. *Statistical Evaluations*
III. *Results and Discussion*
 A. *Synopsis of Results*
 B. *Presentation of Data (Tables, Figures, and Supporting Text)*
 C. *Discussion of Significance*
IV. *Conclusions*

Once you have put real content into such a generic outline, the other essential point in organization is transition from one idea to another.

INAPPROPRIATE CONTENT

Too Much for One Paper

Excess may come with the paper's covering too many diverse points; containing too much material; having too much speculation; being too wordy, redundant, or repetitious; or being overloaded with data that could be expressed in representative samples.

Too Little in One Paper

Scientists are far more likely to put too much in a paper than too little, but a lack of detailed development of specific points is also common.

POOR CONSTRUCTION

Closely related to organization, unity within segments (sections, paragraphs, and sentences) as well as the transitions between units is vital to successful construction of the entire paper. Weaknesses in smaller units of the paper occur both in the presentation of data and in sentence construction.

Data

Failure to achieve reader-friendly tables and figures occurs when they contain too little explanation or too much data. They may be arranged in an unconventional order, have poor or faulty headings and legends, be too complex or cluttered, or be unable to stand alone without the text.

Sentence Construction

Sentences can and should be constructed in a variety of ways. Look at the following sentences; they all say essentially the same thing.

a. *Some strata of the earth contain water.*
b. *Water is present in some strata of the earth.*
c. *Rock and sand strata of the earth may hold water deposits.*
d. *Water has been deposited in the earth's strata.*
e. *There is water in rock and sand strata of the earth.*
f. *Contained within the depths of the earth are extensive strata composed of rock, gravel, or sand, some of which collected large deposits of water billions of years ago and still hold those deposits today.*
g. *Deposits called groundwater exist in rock and sand strata of the earth.*
h. *Underground are deposits of water.*
i. *Hidden in the monstrous recesses of the interior of the earth lie extensive strata of rock and sand wherein there exist enormous volumes of water.*

A dozen other versions of this information could be composed. No single construction is best except in the context in which it appears. Two frequent problems in sentence construction are wordiness and misplaced elements.

Wordiness Wordiness often comes with failure to start with primary ideas. Look at the following example that came from a graduate student:

> *It can be noted that salmonellae are present during all phases of poultry production and processing. Although similar hygiene practices were practiced on all of the 10 poultry farms we examined in this study, great variation existed in the degree of salmonella contamination on them. From the results of this study, it appears that salmonellae may be transmitted continuously through feed to the breeder parent stock, to the chicks, through the processing and finally to the finished broiler product.*

We can communicate the same message with

> *Salmonellae is transmitted progressively from feed to breeder chickens and their offspring and then through the processing plant to the finished product. On 10 poultry farms using similar production practices, we found great differences in the degree of salmonella contamination.*

The two ideas (sentences) in this paragraph need further development, probably in separate paragraphs. It would really not matter which of the sentences was developed first or last. We can get rid of the repetition of such things as *"in this study," "practices were practiced,"* and *"all phases of poultry production and processing."* We can also omit rather meaningless phrases such as *"It can be noted that"* or *"it appears that."* The result is that we have cut the length almost in half, and our two points are more prominently displayed and ready to be developed with supporting details.

Misplaced Elements Misplaced elements in sentence construction interfere with smooth reading or logical sequence of ideas. Again, students have furnished examples:

> *Neither callus tissue from the spinach culture in 1988 nor 1989 produced shoots.*

Logically, the "Neither . . . nor" should be connecting only the years. The sentence could become

> *Callus tissue from the spinach culture produced no shoots in either 1988 or 1989.*

Another student wrote,

> *Our purpose was to determine whether the cultivar was more tolerant than others to the pathogen and to characterize the wilt.*

The sentence reads more smoothly if we put the shorter element closer to the verb and write

> *Our purposes were to characterize the wilt and to determine whether the cultivar was more tolerant than others to the pathogen.*

DISTRACTING LITTLE THINGS

The reader's attention is distracted by small inconsistencies and errors. Be consistent. Is it Figure, figure, Fig, fig, Fig., fig., or *Figure, figure, Fig, fig, Fig.*, or *fig*. The term might even be underlined or written in boldface. Hyphens and numbers are used in a variety of ways: *a 3-cm depth, 3 cm deep, two rates, a 2-mg rate, six plants,* or *42 plants.* Use little elements, such as abbreviations, hyphens, numbers, and capitalization, accurately and consistently.

Errors and inconsistencies in bibliographies and textual citations are distracting and may cause a reader to fail to find a reference. Follow publishers' styles and grammatical standards and be sure textual citations coordinate with the entries in the list of references. Misspellings, faulty punctuation, and grammatical errors or inconsistencies anywhere in the text or bibliography can be frustrating to a reader.

SENSITIVITY TO WORDS (DICTION)

Watch out for words with similar meanings or forms. Use the best one available. Dangers exist with *affect, effect; medium, media; data; different, varying;*

while; only; cheap, economical, inexpensive; there is; a small size or a red color; and numerous other terms. Avoid the use of groups of words that express the same meaning as a single word. For example, *at this point in time* means *now,* and a *long time period* means *a long time.* Proofread carefully, and don't trust a computer spell check entirely. Believe it or not, the following have appeared in papers I have reviewed. I'm sure they were all spellchecked.

> *This system has been wildly applied by most laboratories in the U. S.*
>
> *The amount of plant material . . . was not sadistically different between years.*
>
> *Cotton responds to both soil moisture and relative humility.*
>
> *Magnesium sulfate had a notorious decrease on the value of moisture absorbed.*

Appendix 2
SAMPLE LITERATURE REVIEW

Terry Gentry wrote the following literature review in conjunction with the proposal in Appendix 3. Like any good literature review, it serves (1) as an introduction to the study, (2) as justification for the research, and (3) as background information to make the study more meaningful to the researcher as well as the reader. It introduces the problem (polycyclic aromatic hydrocarbon contamination) and possible solutions to the problem (dissipation, bioremediation) with emphasis on the solution proposed in the study (phytoremediation via microorganisms in the rhizosphere). Organization and content for the review are focused on these primary issues, which provided the key words for the literature search that supplied the content. Note the review is not just a listing or summary of studies done in the area but a smooth, organized discussion of the issues with appropriate citations interspersed.

POLYCYCLIC AROMATIC HYDROCARBON INFLUENCE ON RHIZOSPHERE MICROBIAL ECOLOGY

Contamination of soil by toxic organic chemicals is widespread and frequent. This is sometimes the result of large-scale incidents such as the Exxon Valdez oil spill in Alaska (Pritchard and Costa, 1991). But, more often, smaller areas of soil are polluted. Cole (1992) estimated that in the United States there are 0.5 to 1.5 million underground storage tanks leaking into the surrounding soil. *In situ* bioremediation of these contaminated sites may be more feasible than chemical and physical clean-up methods. Degradation of polycyclic aromatic hydrocarbons (PAHs), a major constituent of many of these pollutants, can be possible if PAH-degrading microorganisms are present at the site. These microorganisms may be more prolific in the rhizosphere of plants than in soil with no vegetation.

A. Polycyclic Aromatic Hydrocarbons (PAHs)

Polycyclic aromatic hydrocarbons are organic compounds that are typically toxic and recalcitrant (Sims and Overcash, 1983). They consist of at least three benzene rings joined in a linear, angular, or cluster array (Cerniglia, 1992). Edwards (1983) described PAHs as being practically insoluble in water. They are produced by various processes including the incomplete combustion of organic compounds such as petroleum (Giger and Blumer, 1974; Laflamme and Hites, 1978). The carcinogenicity of many PAHs has been well documented (Haddow, 1974). This knowledge has prompted much research to determine the mode by which these compounds cause cancer and their ultimate health risks to humans (Miller and Miller, 1981). Due to their toxic nature, the United States Environmental Protection Agency included several PAHs in their list of priority pollutants to be monitored in industrial wastewaters (Keith and Telliard, 1979). Heitkamp and Cerniglia (1988) concluded that this interest has resulted in increased efforts to remediate PAH-contaminated soil.

B. Dissipation

Reilley *et al.* (1996) reported the fate of PAHs in soil includes irreversible sorption, leaching, accumulation by plants, and biodegradation. They also contended that surface adsorption is the main process controlling PAH destination in soil. Many PAHs are strongly adsorbed to soil particles (Knox *et al.*, 1993). Means *et al.* (1980) found the PAHs composed of longer chains and greater masses to be more strongly adsorbed to soil particulate matter. Leaching of PAHs from soil is minimal due their adsorption to soil particles and low water solubility (Reilley *et al.*, 1996). Results indicate that larger PAHs may adsorb onto roots, but translocation from roots to foliar portions of the plants is negligible (Edwards, 1983; Sims and Overcash, 1983). Biodegradation is the main pathway by which dissipation can be enhanced.

C. Bioremediation

Bioremediation manipulates biodegradation processes by using living organisms to reduce or eliminate hazards resulting from accumulation of toxic chemicals and other hazardous wastes. According to Bollag and Bollag (1995), two techniques that may be used in bioremediation are (1) stimulation of the activity of the indigenous microorganisms by the addition of nutrients, regulation of redox conditions, optimization of pH, or augmentation of other conditions to produce an environment more conducive to microbial growth and (2) inoculation of the contaminated sites with microorganisms of specific biotransforming abilities.

1. Indigenous Population Soil contains a large and diverse population of microorganisms (Tiedje, 1994). The indigenous population of these microorganisms has been manipulated to increase biodegradation. *In situ* bioremediation utilizes organisms at the site of pollution to remove contaminants. Often, indigenous organisms from the contaminated area, which may even have adapted to proliferate on the chemical, are utilized to remove the pollutants (Bollag and Bollag, 1995).

Microbial degradation may be enhanced by aeration, irrigation, and application of fertilizers (Lehtomäki and Niemelä, 1975). In Prince William Sound, Alaska, following the Exxon Valdez oil spill, the application of fertilizers increased biodegradation up to threefold (Pritchard and Costa, 1991).

The relative contributions of bacterial and fungal populations to hydrocarbon mineralization may differ based upon contaminant and soil parameters. Anderson and Domsch (1975) studied the degradation of glucose in several soils. They attributed the majority of mineralization to fungi (60–90%) with relatively minor bacterial contribution (10–40%). It is unclear if fungi are also the principal degraders of hydrocarbons (Bossert and Bartha, 1984). Song *et al.* (1986) reported 82% of *n*-hexadecane mineralization was due to bacteria while fungi contributed only 13%. They concluded that bacteria are the primary degraders of *n*-hexadecane in the soil tested, but additional experiments are necessary before the results can be generalized. In a field study utilizing six oils as contaminants, Raymond *et al.* (1976) noted that fungi appeared to be the principal hydrocarbon degraders.

From a review of the literature, Cerniglia (1992) found various bacteria, fungi, and algae reported to degrade PAHs. More specifically, Déziel *et al.* (1996) isolated 23 bacteria capable of utilizing naphthalene and phenanthrene as their sole growth substrate. These bacteria were all fluorescent pseudomonads. Shabad and Cohan (1972) reported that soil bacteria are the primary degraders of PAHs. Cerniglia's (1992) review concluded that the microbial degradation of smaller PAHs such as phenanthrene has been thoroughly investigated; however, there has not been sufficient research on the microorganisms capable of degrading PAHs containing four or more aromatic rings. There remains a need for isolation and identification of microorganisms capable of degrading the more persistent and toxic PAHs (Cerniglia, 1992).

2. Introduced Microorganisms Organisms capable of breaking down certain pollutants are not present at all sites; therefore, inoculation of the soil with microorganisms, or bioaugmentation, is sometimes required (Alexander, 1994). Indigenous or exogenous microorganisms may be applied to the polluted soil (Turco and Sadowsky, 1995). Microorganisms capable of degrading several pollutants including PAHs have been isolated from contaminated soil (Heitkamp and Cerniglia, 1988). In addition, Lindow *et al.* (1989) communicated a need for the continued development of genetically engineered microorganisms including those capable of degrading a variety of pollutants. Nevertheless, successful establishment of introduced microorganisms remains enigmatic (Pritchard, 1992; Turco and Sadowsky, 1995). Thies *et al.* (1991) linked the poor survival of introduced microorganisms to competition from native soil microorganisms.

The characteristics that allow introduced microorganisms to become acclimated to a new environment have not been completely elucidated (Turco and Sadowsky, 1995). However, the indigenous soil populations appear to have specific qualities, such as the ability to utilize a particular growth substrate, that give them a competitive advantage in occupying available niches (Atlas and Bartha, 1993). One way to encourage the growth of introduced microorganisms may be to supply a new niche for microbial growth in the form of a suitable plant.

D. Phytoremediation

Phytoremediation is defined by Cunningham and Lee (1995) as "the use of green plants to remove, contain, or render harmless environmental contaminants." This applies to all plant-influenced biological, microbial, chemical, and physical processes that contribute to the remediation of contaminated sites (Cunningham and Berti, 1993). Plants have historically been developed for food or fiber production. With an increasing interest in the use of plants to reduce contamination from organic chemicals, plants may be selected and developed based upon their suitability for bioremediation. Cunningham and Lee (1995) contend that plant attributes such as rooting depth, structure, and density can be altered to increase biodegradation. They assert that if contaminants are (1) in the upper portion of the soil, (2) resistant to leaching, and (3) not an immediate hazard, many may be removed by phytoremediation. Experiments may confirm that phytoremediation is a less expensive, more permanent, and less invasive technique than many current methods of remediation (Cunningham and Lee, 1995).

E. The Rhizosphere

Curl and Truelove (1986) have described the rhizosphere as the zone of soil under the direct influence of plant roots and in which there is an increased level of microbial numbers and activity. They report that the ratio of bacteria and fungi in the rhizosphere to the non-rhizosphere soil (R/S ratio) commonly ranges from 2 to 20 due to the root exudation of easily metabolizable substrates. These exudates include sugars, amino compounds, organic acids, fatty acids, growth factors, and nucleotides (Curl and Truelove, 1986). Legumes usually demonstrate a greater rhizosphere effect than non-legumes (Atlas and Bartha, 1993). Also, the development of plant roots in previously nonvegetated soil may alter soil environmental conditions including carbon dioxide and oxygen concentrations, osmotic and redox potentials, pH, and moisture content (Anderson and Coats, 1995).

Generally, the rhizosphere is colonized by a predominantly gram-negative bacterial community (Curl and Truelove, 1986). Anderson and Coats (1995) reported that one of the interesting and repeated topics discussed during the 1993 American Chemical Society symposium was the prevalence of gram-negative microorganisms in the rhizosphere. Reportedly, the ability of gram-negative bacteria to quickly metabolize root exudates contributes to their predominance in the rhizosphere (Atlas and Bartha, 1993).

Anderson and Coats (1995) suggest that increased rates of contaminant degradation in the rhizosphere compared to nonvegetated soil may result from increased numbers and diversity of microorganisms.

1. Rhizosphere Effect on PAHs The rhizosphere of numerous plants has been reported to increase the biodegradation of several PAHs. Aprill and Sims (1990) examined the effects of eight prairie grasses (*Andropogon gerardi, Sorghastrum nutans, Panicum virgatum, Elymus canadensis, Schizachyrium scoparius, Bouteloua curtipendula, Agropyron smithii,* and *Bouteloua gracilis*) on the biodegradation of four PAHs, benzo(a)pyrene, benz(a)anthracene, chrysene, and dibenz(a,h)anthracene. They reported significantly greater disappearance of the PAHs in the vegetated soils compared to the unvegetated soils, and the rate of disappearance was directly related to the water solubility of each compound.

Reilley *et al.* (1996) investigated the rhizosphere effect of alfalfa (*Medicago sativa* L.), fescue (*Festuca arundinacea* Schreb.), sudangrass (*Sorghum vulgare* L.), and switchgrass (*Panicum virgatum* L.) on the degradation of pyrene and anthracene. They reported that the vegetation significantly increased the degradation of these PAHs in the soil. They concluded that degradation most likely resulted from an elevated microbial population in the rhizosphere due to the presence of root exudates.

Nichols *et al.* (1996) conducted an experiment on the degradation of a model organic contaminant (MOC) composed of six organic chemicals including two PAHs (phenanthrene and pyrene) in the rhizospheres of alfalfa (*Medicago sativa*, var. Vernal) and alpine bluegrass (*Poa alpina*). They found increased numbers of hydrocarbon-degrading microorganisms in the rhizospheres of both plants. From the same study, Rogers *et al.* (1996) reported that plants demonstrated no significant impact on the degradation of the MOC. They concluded it was probable that biological and/or abiotic processes occurring before plants developed enough to produce a rhizosphere effect were responsible for the disappearance of the MOC compounds.

2. Rhizosphere Microbial Ecology in PAH-Contaminated Soil Before microorganisms can be successfully introduced into the soil or managed for increased bioremediation, an increased understanding of the determinants of rhizosphere microbial ecology needs to be developed. Anderson and Coats (1995) stated the need for an expanded understanding of the interactions between plants, microorganisms, and chemicals in the root zone in order to identify conditions where phytoremediation using rhizosphere microorganisms is most feasible.

To date, no studies have been conducted on rhizosphere microbial ecology in PAH-contaminated soil. Furthermore, little is known about the factors controlling rhizosphere microbial ecology in uncontaminated soil. Bowen (1980) asserted the plant to be the predominant force in the rhizosphere system. In contrast, Bachmann and Kinzel (1992) reported that, in a study involving six plants and four soils, the soil was the dominant factor in some plant–soil combinations. A unique symbiosis that developed from the combination of a specific plant and soil microorganisms was evident. Of all tested plants, *Medicago sativa* had the strongest influence on the soil. They concluded that this effect was consistent with the results of Angers and Mehuys (1990) and may be related to the nitrogen-fixing activity of alfalfa.

Additionally, recent research suggests that gram-positive bacteria may be a larger component of the rhizosphere microbial population than previously reported. Cattelan *et al.* (1995) found a large percentage of soybean (*Glycine max*) rhizosphere population to be occupied by the gram-positive bacterial genus *Bacillus* spp. Also, it appears that gram-positive microorganisms may play a major role in the breakdown of contaminants including PAHs. Heitkamp and Cerniglia (1988) isolated a gram-positive bacterium capable of degrading several PAHs. The bacterium could not utilize PAHs as the sole C source, but it did completely mineralize PAHs when supplied with common organic carbon sources such as peptone and starch. Additional research is needed to elucidate the determinants of rhizosphere microbial ecology especially in PAH-contaminated soils.

References

Alexander, M. (1994). *Biodegradation and Bioremediation.* Academic Press, San Diego.

Anderson, J. P. E., and Domsch, K. H. (1975). Measurement of bacterial and fungal contributions to respiration of selected agricultural and forest soils. *Can. J. Microbiol.* **21,** 314–322.

Anderson, T. A., and Coats, J. R. (1995). An overview of microbial degradation in the rhizosphere and its implications for bioremediation. In *Bioremediation: Science and Applications* (H. D. Skipper and R. F. Turco, Eds.), SSSA Spec. Publ. 43, pp. 135–143. ASA, CSSA, and SSSA, Madison, WI.

Angers, D. A., and Mehuys, G. R. (1990). Barley and alfalfa cropping effects on carbohydrate contents of a clay soil and its size fractions. *Soil Biol. Biochem.* **22,** 285–288.

Aprill, W., and Sims, R. C. (1990). Evaluation of the use of prairie grasses for stimulating polycyclic aromatic hydrocarbon treatment in soil. *Chemosphere* **20,** 253–265.

Atlas, R. M., and Bartha, R. (1993). *Microbial Ecology: Fundamentals and Applications,* 3rd ed. Benjamin/Cummings, Menlo Park, CA.

Bachmann, G., and Kinzel, H. (1992). Physiological and ecological aspects of the interactions between plant roots and rhizosphere soil. *Soil Biol. Biochem.* **24,** 543–552.

Bollag, J.-M., and Bollag, W. B. (1995). Soil contamination and the feasibility of biological remediation. In *Bioremediation: Science and Applications* (H. D. Skipper and R. F. Turco, Eds), SSSA Spec. Publ. 43, pp. 1–12. ASA, CSSA, and SSSA, Madison, WI.

Bossert, I., and Bartha, R. (1984). The fate of petroleum in soil ecosystems. In *Petroleum Microbiology* (R. M. Atlas, Ed.), pp. 435–473. Macmillan, New York.

Bowen, G. D. (1980). Misconceptions, concepts and approaches in rhizosphere biology. In *Contemporary Microbial Ecology* (D. C. Ellwood, J. N. Hedger, M. F. Latham, J. M. Lynch, and J. H. Slater, Eds.), pp. 283–304. Academic Press, New York.

Cattelan, A. J., Hartel, P. G., and Fuhrmann, J. J. (1995). Successional changes of bacteria in the rhizosphere of nodulating and non-nodulating soybean (*Glycine max* (L.) Merr.). 15th North American Conference on Symbiotic Nitrogen Fixation, August 13–17, 1995, Raleigh, NC.

Cerniglia, C. E. (1992). Biodegradation of polycyclic aromatic hydrocarbons. *Biodegradation* **3,** 351–368.

Cole, G. M. (1992). *Underground Storage Tank Installation and Management.* Lewis, Chelsea, MI.

Cunningham, S. D., and Berti, W. R. (1993). The remediation of contaminated soils with green plants: An overview. *In Vitro Cell. Dev. Biol.* **29P,** 207–212.

Cunningham, S. D., and Lee, C. R. (1995). Phytoremediation: Plant-based remediation of contaminated soils and sediments. In *Bioremediation: Science and Applications.* (H. D. Skipper and R. F. Turco, Eds.), SSSA Spec. Publ. 43, pp. 145–156. ASA, CSSA, and SSSA, Madison, WI.

Curl, E. A., and Truelove, B. (1986). *The Rhizosphere.* Springer-Verlag, Berlin.

Déziel, É., Paquette, G., Villemur, R., Lépine, F., and Bisaillon, J.-G. (1996). Biosurfactant production by a soil *Pseudomonas* strain growing on polycyclic aromatic hydrocarbons. *Appl. Environ. Microbiol.* **62,** 1908–1912.

Edwards, N. T. (1983). Polycyclic aromatic hydrocarbons (PAH's) in the terrestrial environment—A review. *J. Environ. Qual.* **12,** 427–441.

Giger, W., and Blumer, M. (1974). Polycyclic aromatic hydrocarbons in the environment: Isolation and characterization by chromatography, visible, ultraviolet, and mass spectrometry. *Anal. Chem.* **46,** 1663–1671.

Haddow, A. (1974). Sir Ernest Laurence Kennaway FRS, 1881–1958: Chemical causation of cancer then and today. *Perspect. Biol. Med.* **17,** 543–588.

Heitkamp, M. A., and Cerniglia, C. E. (1988). Mineralization of polycyclic aromatic hydrocarbons by a bacterium isolated from sediment below an oil field. *Appl. Environ. Microbiol.* **54,** 1612–1614.

Keith, L. H., and Telliard, W. A. (1979). Priority pollutants. I. A perspective view. *Environ. Sci. Technol.* **13,** 416–423.

Knox, R. C., Sabatini, D. A., and Canter, L. W. (1993). *Subsurface Transport and Fate Processes.* Lewis, Boca Raton, FL.

Laflamme, R. E., and Hites, R. A. (1978). The global distribution of polycyclic aromatic hydrocarbons in recent sediments. *Geochim. Cosmochim. Acta* **42,** 289–303.

Lehtomäki, M., and Niemelä, S. (1975). Improving microbial degradation of oil in soil. *Ambio* **4,** 126–129.

Lindow, S. E., Panopoulos, N. J., and McFarland, B. L. (1989). Genetic engineering of bacteria from managed and natural habitats. *Science* **244,** 1300–1307.

Means, J. C., Wood, S. G., Hassett, J. J., and Banwart, W. L. (1980). Sorption of polynuclear aromatic hydrocarbons by sediments and soils. *Environ. Sci. Technol.* **14,** 1524–1528.

Miller, E. C., and Miller, J. A. (1981). Searches for ultimate chemical carcinogens and their reactions with cellular macromolecules. *Cancer* **47,** 2327–2345.

Nichols, T. D., Wolf, D. C., Rogers, H. B., Beyrouty, C. A., and Reynolds, C. M. (1996). Rhizosphere microbial populations in contaminated soils. *Water Air Soil Pollut.*, in press.

Pritchard, P. H. (1992). Use of inoculation in bioremediation. *Curr. Opin. Biotechnol.* **3,** 232–243.

Pritchard, P. F., and Costa, C. T. (1991). EPA's Alaska oil spill bioremediation project. *Environ. Sci. Technol.* **25,** 372–379.

Raymond, R. L., Hudson, J. O, and Jamison, V. W. (1976). Oil degradation in soil. *Appl. Environ. Microbiol.* **31,** 522–535.

Reilley, K. A., Banks, M. K., and Schwab, A. P. (1996). Dissipation of polycyclic aromatic hydrocarbons in the rhizosphere. *J. Environ. Qual.* **25,** 212–219.

Rogers, H. B., Beyrouty, C. A., Nichols, T. D., Wolf, D. C., and Reynolds, C. M. (1996). Selection of cold-tolerant plants for growth in soils contaminated with organics. *J. Soil Contam.* **5,** 171–186.

Shabad, L. M., and Cohan, Y. L. (1972). The contents of benzo(a)pyrene in some crops. *Arch. Geschwultsforsch.* **40,** 237–246.

Sims, R. C., and Overcash, M. R. (1983). Fate of polynuclear aromatic compounds (PNA's) in soil–plant systems. *Residue Rev.* **88,** 1–68.

Song, H.-G., Pedersen, T. A., and Bartha, R. (1986). Hydrocarbon mineralization in soil: Relative bacterial and fungal contribution. *Soil Biol. Biochem.* **18,** 109–111.

Thies, J. E., Singleton, P. W., and Bohlool, B. B. (1991). Influence of the size of indigenous rhizobial populations on establishment and symbiotic performance of introduced rhizobia on field-grown legumes. *Appl. Environ. Microbiol.* **57,** 19–28.

Tiedje, J. M. (1994). Microbial diversity: Of value to whom? *Am. Soc. Microbiol. News* **60,** 524–525.

Turco, R. F., and Sadowsky, M. (1995). The microflora of bioremediation. In *Bioremediation: Science and Applications* (H. D. Skipper and R. F. Turco, Eds.), SSSA Spec. Publ. 43, pp. 87–102. ASA, CSSA, and SSSA, Madison, WI.

Appendix 3
SAMPLE PROPOSAL

The literature review in Appendix 2, the slide set in Appendix 12, and the proposal below are the work of Terry J. Gentry, a candidate for the master's degree at the University of Arkansas—Fayetteville (used with permission of T. J. Gentry). Both the review of literature and the proposal were presented to his committee along with his proposed plan of course work. The slide set was used with a seminar presentation and also presented to his committee with his proposal. In addition, his preliminary work was used as the basis for a poster at a regional meeting.

Notice in this proposal the inclusion of the preliminary work, which is not a part of many proposals. The proposal could have been presented to the graduate committee prior to the preliminary work. In that situation, the proposal would contain two objectives, and conditions of the second objective and methods would depend on results from the first. With inclusion of the preliminary study here, however, Mr. Gentry has additional justification for his proposed study and has already determined the conditions on which the objective and methods will be based. The methods section of his "Proposed Experiment" can simply refer to that section of the "Preliminary Study." Otherwise, if the preliminary study were removed, it would still be a complete, unified proposal. Also, some graduate committees may want the literature review incorporated into the proposal itself. The following model, then, is not to be considered a set standard; individual proposals must fit the situation and the experimentation proposed.

POLYCYCLIC AROMATIC HYDROCARBON INFLUENCE ON RHIZOSPHERE MICROBIAL ECOLOGY

INTRODUCTION

Contamination of soil by toxic organic chemicals is widespread and frequent. This is sometimes the result of large-scale incidents such as the Exxon Valdez oil spill in Alaska (Pritchard and Costa, 1991). But, more often, smaller areas of soil are polluted. Cole (1992) estimated that in the United States there are 0.5 to 1.5 million underground storage tanks leaking into the surrounding soil. Polycyclic aromatic hydrocarbons (PAHs) are a major constituent of many of these pollutants. These PAHs are characterized by their toxicity and persistence (Sims and Overcash, 1983).

Phytoremediation is a technology that may provide cheaper, more permanent, and less invasive amelioration of these contaminated soils than other remediation methods (Cunningham and Lee, 1995). Curl and Truelove (1986) described the rhizosphere as the zone of soil under the direct influence of plant roots and in which there is an increased level of microbial numbers and activity. The rhizosphere of numerous plants has been shown to increase the biodegradation of various organic contaminants (Anderson and Coats, 1995). Aprill and Sims (1990) found increased disappearance of four PAHs in soil columns planted with prairie grass compared with nonvegetated soil columns. Reilley *et al.* (1996) reported an increased degradation of pyrene and anthracene in the rhizospheres of three grasses and a legume.

Although much work has been done on manipulating the soil microflora, little is known about the determinants of rhizosphere microbial ecology (Bachmann and Kinzel, 1992). Anderson and Coats (1995) suggest that a better understanding of the mechanistic interactions among plants, microorganisms, and chemicals in the root zone will help identify situations where phytoremediation can be appropriate.

In a preliminary study we assessed the diversity of microorganisms present in two soils without vegetation and in the rhizosphere of bahiagrass (*Paspalum notatum* Flügge, var. Argentine). The goal of the proposed study is to determine how populations of microorganisms in the rhizosphere respond to the presence of a PAH.

PRELIMINARY STUDY

The objective of the preliminary experiment was to assess the impact of bahiagrass rhizosphere on bacterial density and diversity in two soils.

Materials and Methods

The experiment was conducted under preset conditions on a Captina silt loam (fine-silty, siliceous, mesic Typic Fragiudults) at the University of Arkansas—Fayetteville and on an Appling sandy loam (clayey, kaolinitic, thermic Typic Kanhapludults) at the University of Georgia—Athens. Field moist samples from the Ap horizons of the soils were passed through sterile 2-mm sieves and stored at 4°C until the beginning of the experiment. Cone-tainers® were surface-sterilized by submerging in 30% bleach for 15 min and rinsing with sterile deionized water. Sterile cheesecloth patches were placed in the bottom of cone-tainers®. Sixty grams of moist soil (52.4 g dry weight) was weighed into respective surface-sterilized cone-tainers® and incubated in growth chambers with day/night cycles of 16/8 h and 27/16 ± 1°C. For 2 weeks prior to planting, soils were maintained at 60% of −0.03 MPa (14.2% w/w) soil moisture potential with sterile distilled water.

To reduce plant genetic variability, bahiagrass, an apomictic plant, was selected for the experiment. One gram of seed was weighed into a scintillation vial. Ten milliliters of 95% ethanol was added to the seed and vortexed. Ethanol was removed from seed by pipette. The vial containing ethanol-cleaned seed was placed onto ice in an ice bucket. Ten milliliters of concentrated sulfuric acid was added to the vial. The seed were vortexed a few seconds every minute during the 8-min scarification process. Sulfuric acid was drawn off by pipette, and the seed were rinsed seven times with distilled water.

Scarified seed were surface-sterilized with 10 mL of 30% bleach solution added to the vial. Seed were sterilized for 30 min and were vortexed for a few seconds every 5 min. Bleach solution was drawn off, and 10 mL of sterile 0.01 *M* HCL was added to the vial, vortexed, and allowed to stand for 10 min. A pipette was used to remove the 0.01 *M* HCL from the vial, and the seed were rinsed with sterile distilled water six times. Sterile seed were inserted between moist sheets of sterile filter paper in Petri dishes and placed in an incubator at 30 ± 1°C. After 4 days, five germinated seeds were planted in respective, preincubated cone-tainers® and covered with 1 cm of sterile sand. Cone-tainers® were returned to the growth chambers and maintained at the same conditions as prior to planting. Seedlings were thinned to one per cone-tainer® after emergence. Bulk soil in control cone-tainers® received no seed but was otherwise treated the same.

Plants were harvested 3 weeks after planting. Cone-tainers® were sectioned vertically with a sterile scalpel, and plant shoots were cut off at the soil surface. Soil was emptied onto sterile aluminum foil. Roots were carefully removed from the soil and lightly shaken against a sterile surface to remove loosely attached soil. Roots plus tightly adhering, rhizosphere soil were placed into 99-mL dilution blanks for the 10^{-2} dilutions. One gram of soil from the nonvegetated soil cone-tainers® was utilized for the corresponding 10^{-2} dilutions of nonrhizosphere (bulk) soil. Samples were serially diluted and spread-plated on 1/10-strength trypticase soy agar (0.1× TSA) containing 100 mg cycloheximide/L. Total numbers of aerobic, heterotrophic bacteria were determined after incubation of plates at 28 ± 1°C for 2 days (Wollum, 1994).

Approximately 60 well-separated bacterial colonies were randomly selected from each treatment: Captina bulk soil, Captina rhizosphere, Appling bulk soil, and Appling rhizosphere. Isolates were restreaked twice on 0.1× TSA plates to check for purity before a final transfer to 1× BBL brand TSA (Becton–Dickinson and Co., Cockeysville, MD). After 24 ± 2 h of growth at 28 ± 1°C, about 40 mg of each isolate was extracted for fatty acid methyl ester (FAME) analysis (Sasser, 1990).

Extracts were sent to the University of Delaware and analyzed by gas chromatography with the MIDI system (MIDI, Newark, DE). Fatty acid profiles were compared with aerobic bacteria in the TSBA library from MIS Standard Libraries. A similarity index of ≥0.4 was considered a match for the genus (Kennedy, 1994; Sasser, 1990).

A 2-sample *t* test was used to compare the total bacterial numbers in the bulk soil and rhizosphere for each soil. The distributions of bacterial genera for soil (Captina, Appling)-treatment (bulk soil, rhizosphere) combinations were compared by a chi-squared test for equality of distributions. Where small expected cell counts invalidated the usual chi-squared test, Fisher's exact test was used. The *P* values at ≤0.05 were considered significant.

Results and Discussion

Total numbers of soil bacteria for the bulk soil and bahiagrass rhizosphere of the Captina silt loam were 1.0×10^7 and 1.1×10^7 colony-forming units (CFU)/g of dry soil, respectively, and were not significantly different (Table 1). Similar data were recently reported for bulk soil and alpine bluegrass (*Poa alpinus*) rhizosphere of Captina silt loam (Nichols *et al.*, 1996). The bluegrass demonstrated a slightly larger rhizosphere influence than the bahiagrass on the total number of bacteria, but this difference may have been partly due to the longer growth period utilized in that experiment.

TABLE 1
Total Bacterial Numbers in Bulk Soil and Bahiagrass Rhizosphere (Rhiz) of Captina and Appling Soils

Captina silt loam			Appling sandy loam		
Bulk	Rhiz	R/S*	Bulk	Rhiz	R/S
—	10^6 CFU/g dry soil	—	—	10^6 CFU/g dry soil	—
10.0a**	11.0a	1.1	5.8a	18.3b	3.2

*Ratio of rhizosphere/bulk soil populations.
**For a given soil, numbers with the same lowercase letter are not significantly different at the 5% level.

In contrast, total bacterial numbers of the Appling sandy loam were 0.6×10^7 and 1.8×10^7 CFU/g dry soil in the bulk soil and rhizosphere, respectively, and were significantly different. A comparable increase in bacterial numbers has been shown for a soybean (*Glycine max*) rhizosphere in Appling sandy loam (Cattelan *et al.*, 1995).

The observed rhizosphere influence on bacterial numbers was relatively small with R/S ratios of 1.1 for the Captina silt loam and 3.2 for the Appling sandy loam. Published data indicate that grasses typically exhibit low R/S ratios as compared to legumes (Curl and Truelove, 1986).

Although there was no measurable difference in the total number of bacteria between the bulk soil and bahiagrass rhizosphere of Captina silt loam, there was a significant increase in diversity in the rhizosphere (Fig. 1). The percentage of *Bacillus* spp. decreased from 84% in the bulk soil to 25% in the rhizosphere. Accordingly, the number of bacterial genera represented in the rhizosphere was increased with the appearance of *Alcaligenes* and *Burkholderia* + *Pseudomonas* spp. There was a concomitant increase in the proportion of *Arthrobacter* and *Micrococcus* spp. Also, the number of unidentified bacteria (similarity indices <0.40) increased from 12% in the bulk soil to 39% in the rhizosphere.

Despite the decrease in the percentage of *Bacillus* spp., the gram-positive bacteria *Bacillus* and *Arthrobacter* were dominant in the Captina bulk soil and rhizosphere. Only small numbers of gram-negative bacteria such as *Pseudomonas* were identified. These results differ from those reported by other investigators that gram-negative bacteria predominate in the rhizosphere (Curl and Truelove, 1986). The gram status of unidentified isolates has not been determined.

In contrast, Appling sandy loam demonstrated no significant change in bacterial diversity between the bulk soil and rhizosphere (Fig. 2). Only three bacterial genera were identified in the bulk soil and in the rhizosphere. Identified bacteria were mostly *Bacillus* and *Arthrobacter* spp., and their numbers were relatively consistent at 21% and 10% in the bulk soil and 25% and 10% in the rhizosphere, respectively. The majority of the isolates from the bulk soil and rhizosphere were not identified, 68% and 63%, respectively. Cattelan *et al.* (1995) reported that there was no significant difference in the relative frequency of bacterial genera of the bulk soil and soybean rhizosphere of Appling sandy loam when sampled at 3 and 15 days after planting.

Again, in contrast to the reported data (Curl and Truelove, 1986) the gram-positive genera *Bacillus* and *Arthrobacter* comprised the majority of identified isolates from the bulk soil and rhizosphere of the Appling sandy loam. In addition, the gram status of a large, unidentified percentage of bacterial isolates has not been determined.

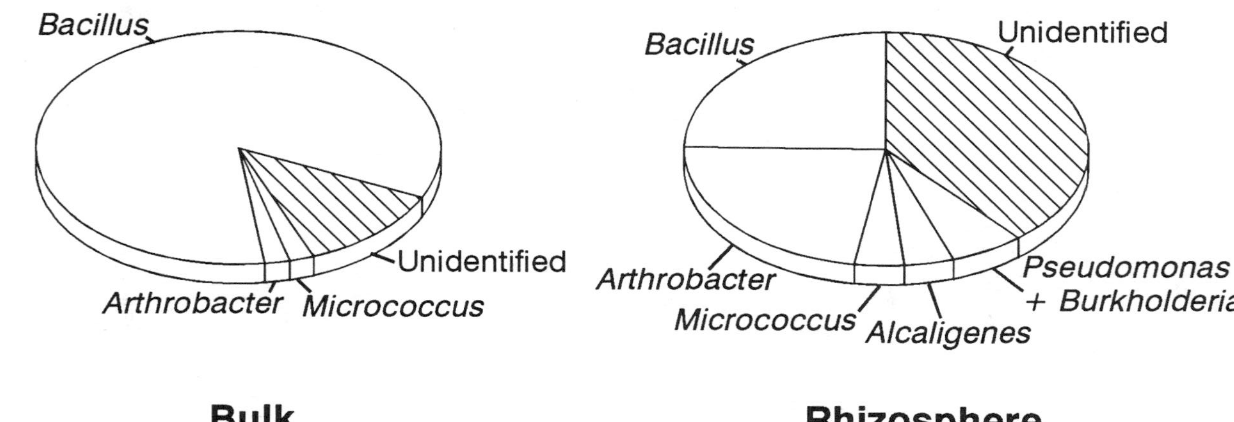

Figure 1. Relative proportions of bacterial genera isolated from the Captina silt loam bulk soil and bahiagrass rhizosphere.

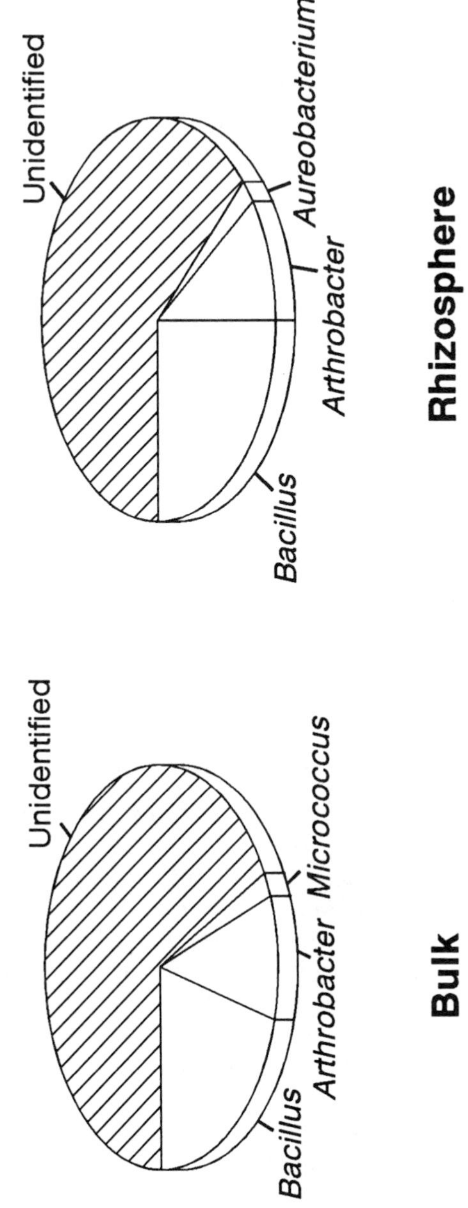

Figure 2. Relative proportions of bacterial genera isolated from the Appling sandy loam bulk soil and bahiagrass rhizosphere.

Conclusions

The data indicate that bahiagrass rhizosphere significantly increased the bacterial diversity in the Captina silt loam but did not influence the bacterial diversity in the Appling sandy loam. The soil appeared to impact the rhizosphere effect on bacterial diversity. More soils need to be tested to clarify these results. The total rhizosphere influence on bacterial numbers and diversity may be underestimated due to the inability to culture and/or identify a large portion of the soil bacterial population by current methods.

PROPOSED EXPERIMENT

With some knowledge of its microbial ecology from the preliminary experiment, Captina silt loam has been chosen as the soil to assess the influence of a PAH on the microbial ecology in the rhizosphere. Likewise, bahiagrass will be one of the plants studied. Alfalfa (*Medicago sativa,* var. Vernal), a legume, has been added to the study because of its demonstrated rhizosphere influence (Nichols *et al.,* 1996). Pyrene, a four-ring PAH, will be the contaminant due to its recalcitrance in soil (Rogers *et al.,* 1996).

Objective

The objective of this study is to determine the effects of pyrene on the rhizosphere microbial ecology of bahiagrass and alfalfa.

Materials and Methods

Treatments will be similar to those used in the preliminary study, and comparisons will be made between bulk soil, bahiagrass rhizosphere, and alfalfa rhizosphere with and without pyrene. The Captina silt loam will be collected and sieved through a sterile 2-mm sieve. Sieved soil will be stored at 4°C in Ziploc® bags until the experiment begins. Soil equivalent to 750 g dry weight will be added to glass 1.5 pint jars, which have been sterilized by rinsing once with 30% bleach solution and six times with water to remove residual bleach before soil is added. Soil in jars will be pre-incubated for 2 weeks at room temperature. Soil is to be maintained at −0.03 MPa soil moisture potential by daily addition of sterile distilled water.

Pyrene will be ground in a mortar and pestle, passed through a 250-μm sieve, and mixed into soil at a level of either 0 mg/kg or 2000 mg/kg. Controls with no pyrene will be used.

Bahiagrass and alfalfa seed will be germinated and planted into respective jars using procedures outlined in the preliminary study.

Plants will be harvested 10 weeks after planting following procedures outlined in the preliminary study. Samples will also be spread-plated onto Martin's medium to determine the number of fungi and onto starch–casein agar to determine the number of actinomycetes (Wollum, 1994).

Approximately 500 colonies each of the actinomycetes, bacteria, and fungi will be isolated from spread-plates and inoculated into Bushnell–Haas broth containing 50 mg/10 mL pyrene to determine whether the microbial isolates are capable of using pyrene as a sole C source and can degrade the compound. Pyrene in the rhizosphere and nonrhizosphere soil will be extracted and respective levels determined by gas chromatographic analysis.

Statistical analysis will be the same as those outlined in the preliminary study.

Conclusions

Much work has been done to determine the rhizosphere effect on the biodegradation of organic contaminants. Studies have attributed the increased degradation of several compounds in the rhizosphere of numerous plants to microorganisms. In contrast, very few studies have examined the effect of contaminants on the microbial ecology of the rhizosphere. An increased understanding of the microbial interactions in the rhizosphere may make possible more precise manipulation of the rhizosphere microbial population for the biodegradation of organic contaminants. These experiments should identify specific microorganisms capable to degrading the PAH pyrene. Additionally, the increased knowledge of rhizosphere microbial ecology in contaminated soil may lead to the selection of appropriate plants capable of stimulating an indigenous or introduced microbial population to enhance the remediation of contaminated sites.

References

Anderson, T. A., and Coats, J. R. (1995). An overview of microbial degradation in the rhizosphere and its implications for bioremediation. In *Bioremediation: Science and Applications* (H. D. Skipper and R. F. Turco, Eds.), SSSA Spec. Publ. 43, pp. 135–143. ASA, CSSA, and SSSA, Madison, WI.

Aprill, W., and Sims, R. C. (1990). Evaluation of the use of prairie grasses for stimulating polycyclic aromatic hydrocarbon treatment in soil. *Chemosphere* **20,** 253–265.

Bachmann, G., and Kinzel, H. (1992). Physiological and ecological aspects of the interactions between plant roots and rhizosphere soil. *Soil Biol. Biochem.* **24,** 543–552.

Cattelan, A. J., Hartel, P. G., and Fuhrmann, J. J. (1995). Successional changes of bacteria in the rhizosphere of nodulating and non-nodulating soybean (*Glycine max* (L.) Merr.) (poster). 15th North American Conference on Symbiotic Nitrogen Fixation, August 13–17, 1995. Raleigh, NC.

Cole, G. M. (1992). *Underground Storage Tank Installation and Management.* Lewis, Chelsea, MI.

Cunningham, S. D., and Lee, C. R. (1995). Phytoremediation: Plant-based remediation of contaminated soils and sediments. In *Bioremediation: Science and Applications* (H. D. Skipper and R. F. Turco, Eds.), SSSA Spec. Publ. 43, pp. 145–156. ASA, CSSA, and SSSA, Madison, WI.

Curl, E. A., and Truelove, B. (1986). *The Rhizosphere.* Springer-Verlag, Berlin.

Kennedy, A. C. (1994). Carbon utilization and fatty acid profiles for characterization of bacteria. In *Methods of Soil Analysis.* Part 2: *Microbiological and Biochemical Properties* (R. W. Weaver, Ed.) SSSA Book series #5, pp. 543–556. SSSA, Madison, WI.

Nichols, T. D., Wolf, D. C., Rogers, H. B., Beyrouty, C. A., and Reynolds, C. M. (1996). Rhizosphere microbial populations in contaminated soils. *Water Air Soil Pollut.*, in press.

Pritchard, P. H., and Costa, C. T. (1991). EPA's Alaska oil spill bioremediation project. *Environ. Sci. Technol.* **25,** 372–379.

Reilley, K. A., Banks, M. K., and Schwab, A. P. (1996). Dissipation of polycyclic aromatic hydrocarbons in the rhizosphere. *J. Environ. Qual* **25,** 212–219.

Rogers, H. B., Beyrouty, C. A., Nichols, T. D., Wolf, D. C., and Reynolds, C. M. (1996). Selection of cold-tolerant plants for growth in soils contaminated with organics. *J. Soil Contam.* **5,** 171–186.

Sasser, M. (1990). MIDI Technical Note 101: Identification of bacteria by gas chromatography of cellular fatty acids. Microbial ID, Newark, DE.

Sims, R. C., and Overcash, M. R. (1983). Fate of polynuclear aromatic compounds (PNAs) in soil–plant systems. *Residue Rev.* **88,** 1–68.

Wollum, A. G., II (1994). Soil sampling for microbial analysis. In *Methods of Soil Analysis.* Part 2: Microbiological and Biochemical Properties. (R. W. Weaver, Ed.), SSSA Book series #5, pp. 1–14. SSSA, Madison, WI.

Appendix 4
ALTERNATE ROUTES TO THE THESIS

Whether you write your thesis in the traditional format or with publishable journal articles, your ultimate goals are a degree, a publication, and a job or additional education. The thesis itself is only a part of the work that you must incorporate into your plans for the graduate program. Any one of a variety of plans can lead to your success with the thesis, and the road map below simply represents two examples. For simplicity, the routes are assuming a master's thesis that would take approximately 2 to 2.5 years to complete and would result in the publication of one journal article. A more complex doctoral dissertation or additional publications would extend the time and add to the work load along the way. Certainly, the order of events may be rearranged, especially toward the last when the thesis, the defense, the job, and the publication may all close in on you at once. Be careful about entering a new degree program or taking a job before your thesis is complete (Fig. A4-1).

You may feel that you are jumping through a series of hurdles as you move along your road to the thesis. For convenience in discussing the steps along the way, I'm dividing the process arbitrarily into seven hurdles.

HURDLE 1

For your satisfaction and success, the first steps along the road must produce an inviting but practical plan. You must choose a topic for research and an advisor that you can believe in and work with intensively for at least 2 years. You will probably investigate the graduate school you want to attend, study which areas of research are prominent in the department where you wish to

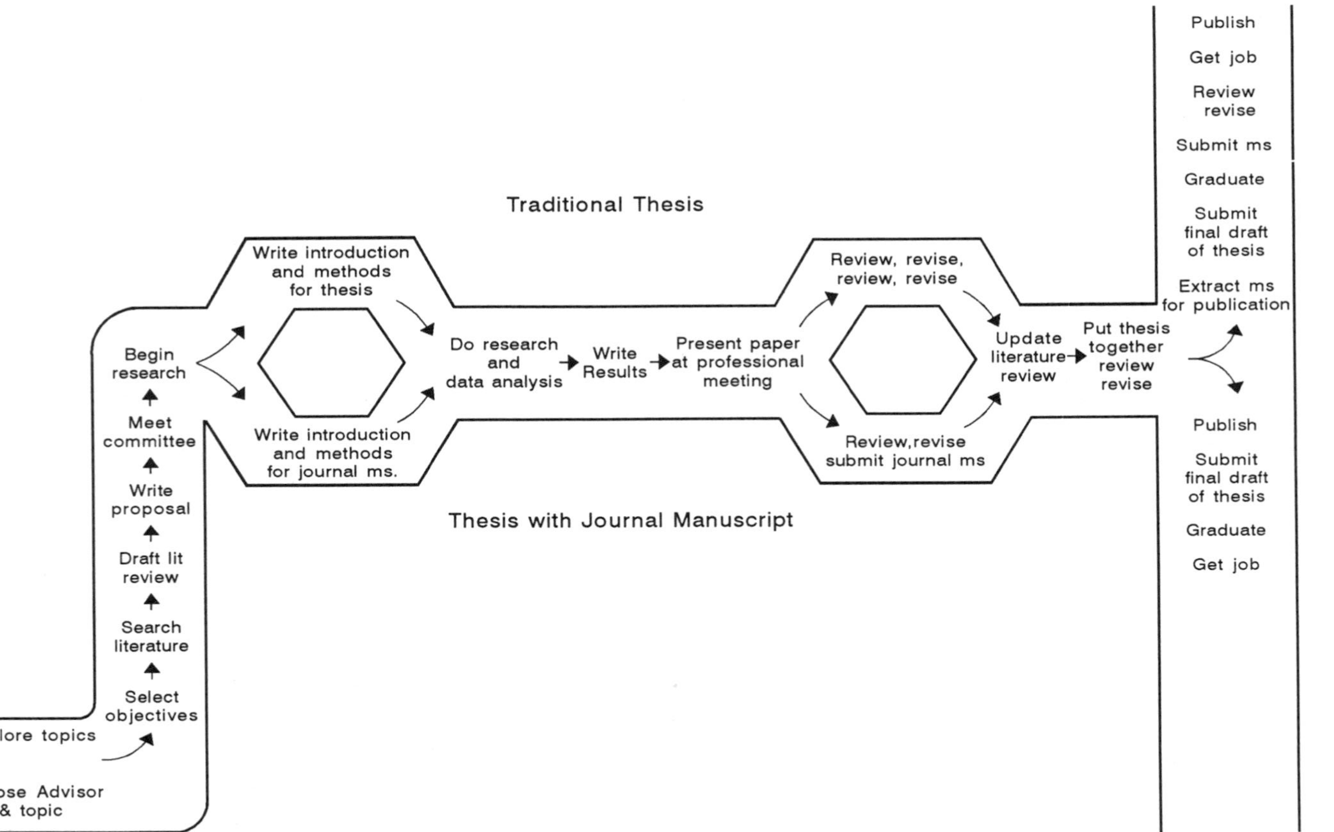
Explore topics
Choose Advisor & topic
Select objectives
Search literature
Draft lit review
Write proposal
Meet committee
Begin research
Traditional Thesis
Write introduction and methods for thesis
Write introduction and methods for journal ms.
Do research and data analysis
Write Results
Present paper at professional meeting
Review, revise, review, revise
Review,revise submit journal ms
Update literature review
Put thesis together review revise
Thesis with Journal Manuscript
Publish
Get job
Review revise
Submit ms
Graduate
Submit final draft of thesis
Extract ms for publication
Publish
Submit final draft of thesis
Graduate
Get job

FIGURE A4-1

enroll, and find out as much as you can about major professors working in your area. Read some of the publications by prospective advisors, and then communicate with them, go through the admissions process, and select or accept a specific advisor.

Through library searches, you should explore topics that are related to the work of the professor you will work with, and when you meet with him or her, have some ideas in mind about what research project you would like to pursue. With your advisor's help, then, you will consider possible hypotheses and select specific objectives for your research. At this point, having read some literature on the subject will be a real help. On the basis of what your subject involves and with the help of your advisor, choose graduate committee members and visit with each of them about your plans. Now you are ready to go to work.

HURDLE 2

With your objectives in mind, search the literature and write a draft of your literature review. This review will be updated toward the end of your program probably to be used as a chapter of your thesis. With that rough draft behind you, write a draft of your proposal, review it, revise it, give it to your major professor for review, and revise again. With approval from your major professor, submit the proposal to your graduate committee and meet with them to discuss your program. Six months may have passed by now. (It will be a year or more if you don't move quickly but meticulously over these first two hurdles.)

HURDLE 3

Except for classes, research will take most of your waking hours for the next year. But don't make your last year harder or extend the time for your program by failing to write along the way. As you begin the research, you should choose which thesis path you will take. The traditional route means your thesis will ultimately consist of an introduction and chapters on literature review, methods, results, and discussion. You may have other addenda such as appendices and an abstract. This traditional thesis, as written, is seldom ready to publish when it is completed, but representative ideas and data can be extracted for publication later. The journal manuscript format means you write with publication in mind from the first, and a literature review chapter and other addenda will hold everything except the representative publishable material. To jump this third hurdle then, you need to write an introduction and a methods chapter for the traditional thesis or an introduction and methods section for the journal manuscript.

HURDLE 4

The research and especially the analysis of data continue, and as they near completion, you write a draft of the results. You are probably at least a year and a half into your program. With these results written, you are ready to go to a professional meeting and make a poster or slide presentation. It is time to think about prospective employers and to start making your reputation public. Continue to review and revise the various sections of the thesis. If you are writing a publishable journal article as a part of your thesis, it should by now be reviewed in-house and ready to submit to the journal.

HURDLE 5

Now get serious about finishing the thesis. For either format, supply the appendices, the abstract, and any other addenda you need. Update the literature review that you wrote over a year ago. Review and revise each section repeatedly, without and with advice from your major professor. Put the thesis together. Review and revise again. It is not unusual to review and revise your thesis six or more times.

HURDLE 6

The paths for the two thesis plans part again at this point. You are over Hurdle 6 when you are ready to defend your thesis. If you are on the journal manuscript route, your manuscript should be in journal review or back to you and ready for subsequent revisions and publication. Be sure that complementary chapters or appendices of your thesis include all the data from your project that are not appropriate for your publication. If you have written a traditional thesis, you should extract representative data, write a journal manuscript, and ask for in-house reviews. Getting a draft of the manuscript ready at this point will make life less miserable as you are seeking employment and beginning a new job. Note that, although the road behind you may have been smoother or simpler than that for your colleague who was writing the journal manuscript all along, once you submit the final draft of the thesis to committee, Hurdle 7 may be somewhat harder to scale with the publication still to be submitted, reviewed, and revised.

HURDLE 7

Reviewing, revising, publishing, defending the thesis, taking exams, presenting a departmental seminar, and applying and interviewing for a job—all may confront you at once. But over this final hurdle lie your three ultimate goals: your degree, a publication, and a job or a new degree program. The order in which the goals are accomplished can depend on your priorities and on which route you took to this point. The publication may be behind you if you took the journal article route. As you approach graduation, establish priorities carefully. The job you want may become available before you finish the thesis. Do not sacrifice the degree prematurely for new short-term goals. The completed degree and a publication can be important stepping stones to your long-range plans. Remember that it is difficult to finish the thesis and produce a publication while you are becoming oriented to a new job. If you are going on for another degree, you may find that completing the thesis and a publication can interfere with scaling Hurdle 1 again. Plan your strategy carefully and make a smooth jump over Hurdle 7 so that reaching your final goals is a satisfying benchmark in your career.

Appendix 5
SAMPLE REVIEW OF MANUSCRIPT SUBMITTED FOR PUBLICATION

The following is a review by Dr. Gail Vander Stoep of Michigan State University that was done for a journal editor on a manuscript submitted for publication (used with permission from Dr. Vander Stoep). It has been abridged for inclusion in this appendix. The marginal notes are mine. Note how thorough and organized the review is with comments on both general areas of content and organization as well as specific points. The reviewer accompanied these remarks with a copy of the manuscript marked clearly with even more specific details that needed to be addressed in the revision. Throughout, there is a tone of concern for the quality of the paper and a recognition of its value. Although this version of the paper was weak, note the positive comments made and the reviewer's ability to criticize and evaluate without being disparaging. The criticism is compelling but is always accompanied by suggestions for amending weaknesses. The authors of the paper expressed sincere appreciation for this review, and by incorporating the suggestions, they produced a good paper, which was subsequently revised twice and then published.

Review of
University-Based Education and Training Programs in Ecotourism or Nature-Based Tourism in the USA
for
Journal of Natural Resources and Life Sciences Education

GENERAL REVIEW, RELEVANCE AND RECOMMENDATION

Reviewer begins with positive comments.

Conscious of the audience of the journal

It appears that the content of the article fits within the scope of the journal content. Issues of resource preservation, conservation, resource use and management, and tourism are not new, but consideration of all of these issues certainly is gaining wider attention across many sectors. University-based education that facilitates systems-based resource analysis, planning and management is critical. I would support publication of this article. However, I recommend revisions to clarify, focus and tighten the paper, as well as to make it more relevant to JNRLSE readers.

Improving relevance for readers:

—Somewhere in the introductory and/or historical perspective sections, indicate (briefly) how ecotourism/nature-based tourism may have implications for various "natural resource" segments (e.g., natural and archaeological resources, . . . links to agriculture-based tourism).
—Expand the "conclusions and recommendations" section, including discussion of relevance to readers, future curriculum development or revision, and the possible need for continuing education.

Title

Specific suggestion with support for her opinion

Consider removing "and Training Programs" from the title because it appears that the scope of this study is education-based and describes individual courses.

General Editorial Comments

Reviewer is flexible and concedes that she may not have all the *best* answers.

I am returning my copy of the paper with rather extensive comments written directly on the manuscript. Many of the notations are strictly editorial and deal with punctuation, capitalization, syntax, grammar, simplification of sentences, or alternative word choices. In some cases the authors may develop something even better or more in line with their thoughts.

Introduction

Another positive statement

I like the introduction as a way to introduce the concept of and need for incorporating ecotourism concepts into university curricula. The introduction also needs to clearly define the need or rationale for the study and its results. WHY do we need to know about ecotourism/nature-based tourism curricula? Clarify why it's a problem that no data are available. (As stated, it's simply "unfortunate," as viewed by the authors.)

Historical Perspective

I believe the historical perspective section is pertinent to the paper, particularly in defining the terms ecotourism and nature-based tourism. However, the discussion could be expanded to address issues of. . . . [*Reviewer goes into detail here on issues that should be addressed.*]

Reviewer clearly rejects this discussion but has positive suggestions for its use in another publication.

I would delete the entire discussion (and Table 1) about ROS and a potential parallel with a TOS. The ideas and linkages are not fully developed; there are some "challengeable" holes in the descriptions. . . . [*Reviewer gives specific examples.*] There is not enough space in this article to fully and logically develop the concept. (Perhaps the idea could be more fully developed for another paper for a different outlet.)

Makes a point on unity and organization

Page 4, starting with line 19: This section seems to be more appropriate as part of the "methods" section than "historical perspective."

Research Methods

Reviewer presents ideas in the form of questions and leaves to the authors the responsibility for determining whether and how to answer them.

Perhaps a little more development of the methods section would be helpful. Some questions I had:
—Any overlap in listing of specific programs/departments among the three lists?
—What types of questions were included in the one-page questionnaire?
—How was content analysis done?
—[*The reviewer continues with a dozen other questions.*]

Results

Points out the need to clarify the meaning of results

Before beginning discussion of actual results, clearly present the response rate versus those responding who OFFER or PLAN TO OFFER a course with ecotourism content. Page 5, lines 16–19 are confusing.
Comment (page 5, lines 21–22) seems to be more of a conclusion than result.

And the need for details

Throughout the "results," clarify what the percentages are based upon; e.g., total number of returned questionnaires, total number of schools reporting current and/or future offering of courses, total number of schools currently offering. . . .

A couple of logic concerns:

Reviewer points to specific parts of the paper and, without revising, helps to direct the logical thinking needed to revise.

—page 8, lines 16–18 (see comment on MS): need transition or link to the list of "other common objectives."
—page 9, lines 8–11, and Figure 5: There seem to be different ways of interpreting "the primary focus" . . . actually as "the primary focus" versus simply "a component of." This is confusing.

Points to specific details that could mislead a reader

Your use of the word "classes" is confusing. Does "class" refer to the four categories or *classifications* of approaches to instruction? or to *courses*? It may be that the word is used interchangeably. . . .

Positive statement

Indirect suggestions for the authors to consider

The descriptions of the four general approaches to instruction are helpful and interesting (and necessary). However, this section could be strengthened by describing HOW the four approaches were determined (perhaps this is more appropriate for the methods section?) and by indicating the number of percentages of courses in this study that use each of the approaches.

Conclusions and Recommendations

This section was deservedly the most criticized part of the paper by all reviewers. This reviewer is direct but has prefaced condemnation of this part with earlier positive comments about the paper's value. She follows her direct remarks on weaknesses with constructive criticism on how to improve the conclusions.

This section is the weakest part of the paper. Most of the elements currently included in the section would seem to me more appropriate elsewhere:
—limitation of scope and diversity: methods section. (However, a recommendation for future research and inventory . . . is appropriate here.)
—Comments about the Ecotourism Society seem more appropriate as a sidebar or footnote. Beginning on line 3, page 13: It is not at all clear how results of the study indicate a need to improve information exchange. This may be a recommendation of the authors but is not directly linked to the results as presented. This section could be strengthened by expanding the conclusions, adding discussion of implications for ecotourism education/training/industry, and by providing specific recommendations. Explain WHY information exchange is important.

Figures and Tables

Reviewer brings up an important point but stops short of encroaching on the editor's territory.

I suspect that some of the figures may not reproduce well, especially if reduced to one-column size. (I defer to the editors for this.) Style guidelines for format (including capitalization) should be checked, then used consistently.

Deals with clarity and content in the tables and figures

Discrepancies exist between results in Figure 2 and the "Year" column of Table 2. (Is Figure 2 needed?)
Figure 4: What were criteria for. . . .
Other comments and questions are written directly on the manuscript. Double check consistency of information and assumptions between tables/figures and narrative within the manuscript.

A final candid, direct opinion on the value of the paper

I recommend that the manuscript be revised with particular attention to clarifying the manuscript and developing a strong "implications" or "relevance" component. It could then be appropriate for publication.

Appendix 6
EVOLUTION OF A TITLE

Suppose that the following six titles are meant to describe the same study. Some of them are better than others. None of them may be fitting for a specific manuscript. Your choice will depend on which words are most important and which best describe the study involved. Think in terms of the words and the arrangement of words that will lead a reader to the central points in your study. The publisher may also have certain criteria for titles such as length and the use of scientific names.

Sample titles:

1. **Controlling the Bollworm**
2. **Investigations into the Effects of Several Selected Phenolic Acid Compounds on the Mortality Rate, Developmental Time, and Pupal Weight Gain of the Cotton (*Gossipium hirsutum* L.) Bollworm (*Helicoverpa zea* Boddie) in Studies Involving Larvae Fed a Synthetic Diet in the Laboratory**
3. **The Effects of Selected Phenolic Compounds on the Mortality, Developmental Time, and Pupal Weight of *Helicoverpa zea* Boddie: Synthetic Diet Studies**
4. **Benzoic and Cinnamic Acids in Synthetic Diets Retard Development of *Helicoverpa zea* Larvae**
5. **Influence of Benzoic and Cinnamic Acids on Mortality or Growth of Bollworm Larvae**
6. **Response of *Helicoverpa zea* Larvae to Benzoic and Cinnamic Acids**

Number 1 might serve as a headline for an article in a newsletter for cotton producers, but it contains too little information to describe a scientific study.

Number 2 is too long and the inclusion of all these words cannot be justified. Certainly, the first three and the last three words can be omitted. Then why not "selected" rather than "several selected"? Why not "phenolic acids" rather than "phenolic acid compounds"? Why not "mortality" rather than "mortality rate"? Why not "pupal weight" rather than "pupal weight gain"? And can't we simply say "bollworm larvae" instead of "bollworm in studies involving larvae"? There is probably no need for the scientific name for cotton because we are naming the cotton bollworm and not the cotton plant. Whether the scientific name *Helicoverpa zea* Boddie appears in a title will depend on the style of the publisher. The authority Boddie might be omitted. Or the scientific names might be used and the common name cotton bollworm might be omitted. These choices would be based on the style of the publisher and the importance of the words for the audience.

Number 3 is also rather long. We might omit "The effects of" and "the" before "mortality," but we still have a long title and must make some other choices. Can we save the "synthetic diet studies" for the abstract? Some publishers avoid two-part titles with the colon. Can we say "development" rather than "developmental time"? Or can we combine the words "developmental time and pupal weight" into a simple term like "growth"? Again, answers to these questions depend on what we need to best describe the study and which key words will allow the readers to retrieve a publication relevant to their interests.

Number 4 is perhaps short enough but could still be improved with the omission of "in synthetic diets" unless that information is vital to a brief description of the study. This title adds a detail by naming the specific phenolics used. This information may be worth the extra space needed. However, the title breaks a convention in scientific writing by using an active verb "retards" that describes the outcome of the study. Characteristically, popular press uses such verbs in headlines, but the scientific report simply describes results of a research effort and discusses outcomes but allows the reader to decide on any final conclusion that can be drawn from the work. What happens under the controlled conditions of a given experiment may not constitute a universal truth, and the active verb appears to be proclaiming such a truth.

Number 5 is about the same length and is similar to number 4 except that it avoids the active verb and uses the common name of the species rather than the scientific name. The choice of name would depend on which one you and your publisher believe will best communicate the information with your audience. The use of "mortality or growth" is somewhat more specific than "development" in number 4 and may be worthy of the extra two words especially if we can get rid of "in synthetic diets."

Number 6 is less descriptive of the paper's content but conserves words. Here I reserve the mortality and growth for the abstract and generalize with the word *Response.* Such a title may be the best choice especially for display on a poster or slide.

Appendix 7

EVOLUTION OF AN ABSTRACT

The following abstract is based on studies done in the laboratory of Dr. Justin R. Morris at the University of Arkansas and is used with his permission. An original abstract from that research has been altered and lengthened with fictitious ideas and data and then pared to acceptable lengths. The examples below show four versions of the same information. Notice changes in organization, content, and wording between the drafts. Little content is lost with a reduction from 373 words to 281 words and then to 218 words in a third version. When I submit this third version with the manuscript to a journal, the editor may inform me that the abstract must be cut to no more than 150 words. To satisfy this request, I have to get rid of the justification and conclusion and reduce methods and results to a bare minimum. The fourth version with 140 words has lost some content but not the most important methods and results. Depending on the editor's preference, the third or fourth version might be the one published.

WORKING ABSTRACT 1 (373 WORDS)

Evaluation of Winegrapes for Suitability in Juice Production

Indirect statement of objective

Abstract. A series of chemical and sensory analyses was designed to determine which, if any, of 10 winegrapes grown in 1994 in Arkansas were suitable for non-alcoholic grape juice production.

Materials

Five of the 10 were classified as red grapes: Chancellor, Cabernet Sauvignon, Villard noir, Cynthiana, and Noble. Five were classified as white grapes: Aurore, Cayuga, Chardonnay, Vidal, and Verdelet. The traditional juice grapes Niagara and Concord were used as controls for comparisons in the study.

Rationale or justification for the study

Sensory quality and consumer acceptability of grape juices depend to a great extent on the process by which the juice is produced but also on the cultivar or blend of cultivars used. With today's processing techniques, winegrape cultivars may also produce non-

Methods

Note the wordiness and repetition throughout, as in "stored at 37 C" and "storage at 37 C."

Results

Also wordy

No overall conclusion

alcoholic juices acceptable to consumers' tastes. Four different means of juice production were used to process the grapes: immediate press, heat process (60 C), heat process (80 C), and 24-h skin contact after pressing. Processed juices were sealed in 0.8-liter bottles and stored at 37 C. The juices were evaluated 1 month after processing and again after 5 months storage at 37 C. Chemical and sensory analyses were run. Chemical analysis showed that red grapes had more acidity than white grapes, but white grapes, except for Cayuga, had more soluble solids. Soluble solid to acid ratios were highest in the red grapes Noble and Cynthiana and lowest in the white Chardonnay and Vidal. Other cultivars showed no significant difference in soluble-solid-to-acid ratios. Chemical analysis showed no difference within cultivar for the treatment process, except that in 6 of the 10 cultivars the 24-h skin contact produced more soluble solids. Consumers' rank preference, as represented by a sensory panel, for flavor of juices revealed greater preference for non-heat treatments, while the heat treatments were more preferred in color evaluations. Flavor was considered to be the most important attribute to the consumer; panelists tended to prefer those juices that had relatively higher soluble-solids-to-acid ratios. The most preferred white grape juices were the immediate press of Niagara and Aurore and the 24-h skin contact treatment of Niagara, Verdelet and Vidal. Rank preference for flavor of red grape juices did not indicate a significant preference among cultivars, thus suggesting all red winegrape cultivars were equally suited for varietal juice production.

WORKING ABSTRACT 2 (281 WORDS)

Title shortened

Rationale shorter and moved to the beginning

Direct statement of objective

Materials

Methods

Results

Less wordiness here than in the first version. Most sentences have been shortened but content is not lost.

Suitability of Winegrapes for Juice Production

Abstract. Recently developed processing methods may provide non-alcoholic juices from winegrape cultivars that are acceptable to consumers. Our objective was to test 10 winegrape cultivars and four juice processes to determine their suitability for juice production. The five red grapes (Chancellor, Cabernet Sauvignon, Villard noir, Cynthiana, and Noble) and five white cultivars (Aurore, Cayuga, Chardonnay, Vidal, and Verdelet) were compared with traditional juice cultivars Niagara (white) and Concord (red). Juices from four processes—immediate press, heat process (60 C), heat process (80 C), and 24-h skin contact after pressing—were sealed in 0.8-liter bottles and stored at 37 C. They were evaluated at 1 month and 5 months after processing. Chemical analysis showed that red grapes had more acidity, but white grapes, except for Cayuga, had more soluble solids. Soluble-solid-to-acid ratios were highest in the red grapes Noble and Cynthiana and lowest in the white Chardonnay and Vidal. Other cultivars showed no significant difference in soluble-solid-to-acid ratios. Chemical analysis showed no difference within cultivar for the treatment processes, except that, in 6 of the 10 cultivars, the 24-h skin contact produced more soluble solids. A sensory panel preferred flavor of juices from non-heat treatments but ranked color best in the heat treatments. Flavor, the most important attribute for consumers, was rated highest in juices that had relatively higher soluble-solids-to-acid ratios.

The most preferred white grape juices were the immediate press of Niagara and Aurore and the 24-h skin contact treatment of Niagara, Verdelet and Vidal. Rank preference for flavor of red grape juices did not indicate a significant preference among cultivars. Most of these winegrapes may be as suitable as juice grapes for non-alcoholic juice production.

Overall conclusion

ABSTRACT, VERSION 3 (218 WORDS)

Suitability of Winegrapes for Juice Production

Rationale shortened even more

Objective

Materials

Methods

Results

Notice that wording is still more concise than in version #2, but little or no content is lost.

A more specific conclusion

Abstract. Recent processing methods may provide acceptable non-alcoholic juices from winegrapes. To determine acceptability of the juices, we compared five red (Chancellor, Cabernet Sauvignon, Villard noir, Cynthiana, and Noble) and five white (Aurore, Cayuga, Chardonnay, Vidal, and Verdelet) winegrapes with traditional juice cultivars Concord (red) and Niagara (white). Juices processed by immediate press, heat process (60 C), heat process (80 C), and 24-h skin contact after pressing were sealed in 0.8-liter bottles and evaluated after 1 month and 5 months storage at 37 C. Chemical analyses showed more acidity in red juices, but more soluble solids (SS) from white grapes except Cayuga. The SS/acid ratios were highest from red grapes Noble and Cynthiana and lowest for white Chardonnay and Vidal with no significant differences in other cultivars. Processes produced no differences within cultivar except the 24-h skin contact produced more SS from six cultivars. A sensory panel ranked color best in the heat treatments but preferred flavor of juices from non-heat treatments. Juices with relatively higher SS/acid ratios rated highest for flavor. Acceptable white grape juices were from the immediate press of Niagara and Aurore and the 24-hr skin contact treatment of Niagara, Verdelet, and Vidal. Panelists indicated acceptance but no flavor preference among red juices. All red winegrapes tested and Aurore, Verdelet, and Vidal white winegrapes appeared suitable for juice production.

ABSTRACT, VERSION 4 (140 WORDS)

Suitability of Winegrapes for Juice Production

No rationale. Objectives and materials combined - specific details are gone.

Basic methods still clear

Abstract. Five red and five white winegrape cultivars were compared with traditional juice grapes Concord and Niagara to determine acceptability of juices. Juices processed by immediate press, heat process (60 C), heat process (80 C), and 24-h skin contact were sealed in 0.8-liter bottles and evaluated after 1 and 5 months storage at 37 C. Red juices had more acidity, but

Results - Here we lose some details contained in version #3, but the most notable results are preserved.

white grapes, except Cayuga, produced more soluble solids (SS). The SS/acid ratios were highest from red grapes Noble and Cynthiana and lowest for white Chardonnay and Vidal. A sensory panel preferred color from heat treatments but flavor of juices from non-heat treatments. Juices with high SS/acid ratios rated highest for flavor. White grape juices were acceptable from immediate press of Niagara and Aurore and 24-h skin contact treatment of Niagara, Verdelet, and Vidal. Panelists indicated acceptance but no flavor preference among red juices.

No conclusion

Appendix 8
PUTTING DATA INTO TABLES AND FIGURES

TABLES

Two versions of the same information appear in the tables on the following page. Notice how Table 1b has refined the information for more clarity and ease in reading.

The arrangement in Table 1a does not allow us to read down from the headings to the information in the stub and field. Reps, or replications, listed in the stub are not germination as their heading states, and numbers in the field are not treatments but germination, and the reader is not told that the measure is in percentages. The 1979 should appear parallel to 1980 and not as part of the main stub head. Notice that the treatment times appear as mixed units of minutes and hours. In addition to these points that cloud the clarity in presentation, too much data may distract from the point made in the table. The analysis has not been done between replications but on combined replications and years. So many numbers may distract from the point made by comparison of the means.

Table 1b improves on the clarity of information by making the table read down, combining the replications as was done in the statistical analysis, using uniform units for measuring time (h) and providing meaning to the numbers in the field with the percentage sign (%). Depending on the importance of this information to the study and space available in the paper, we might use Table 1b in a publication, or we might simply provide the mean percentages and a comment on significant differences in the text without using a table at all.

TABLE 1A

Percentage Germination of *Phytolacca americana* Seeds with Hot Water and Sulfuric Acid Treatments

Germination	Treatments				
		Hot water		Sulfuric acid	
1979	None	90 min/80C	12 h/60 C	15 min	30 min
Rep 1	2	19	30	76	49
Rep 2	0	23	30	54	41
Avg. 2 reps	2	21	30	65	45
1980					
Rep 1	3	42	36	80	42
Rep 2	5	28	32	62	76
Avg. 2 reps	4	35	34	71	59
*Mean (both yr)	3	28c	32c	68c	52b

*Means followed by the same letter are not significantly different at the 0.05 level.

TABLE 1B

Percentage Germination of *Phytolacca americana* Seeds with Hot Water and Sulfuric Acid Treatments

Treatment (time/temp.)	Germination (%) 1979	1980	Mean*
Control	2	4	3d
Hot water (h/C)			
1.5/80	21	35	28c
12.0/60	30	34	32c
Sulfuric acid (h)			
0.25	65	71	68a
0.50	45	59	52b

*Means followed by the same letter are not significantly different at the 0.05 level.

FIGURES

The following figures depict the same fictitious data. In Fig. 1a three of the lines, those for the poultry manures, are not statistically different. The symbols for all six lines in Fig. 1a are black circles, squares, and triangles, and their use with the poultry manure data does not help much in distinguishing one line from another. The lines themselves should be the most prominent images, but the line around the box is heavier and distracting. The entire figure is compressed vertically so that the intervals on the X axis are unconventionally shorter than those on the Y axis. The tic marks need not extend both within and outside the axis lines. The X axis also contains too many numbers, and they do not represent the actual days on which the data points fall. That axis should not be extended beyond the last data point. Space is also wasted in the area between 10^0 and 10^2, and since the Y axis is on a logarithmic scale, it need not begin at 0. Statistically, the fit of the lines with data is not apparent.

Notice the improvements in Fig. 1b. The data for all three kinds of poultry manure have been combined. In the text I will tell my audience what kinds I used and that no significant difference in die-off of *E. coli* occurred among the three. I will make clear that the R^2 represents the fit of the regression line of the geometric means for the three poultry manures. Revisions in the lines, symbols, and axes of Fig. 1b to achieve greater simplicity also make the message conveyed by the data easier to comprehend.

Fig. 1a. *Escherichia coli* die-off in a Captina silt loam amended with cornish hen manure, turkey manure, caged layer manure, beef cattle manure, swine lagoon effluent, or septic tank effluent.

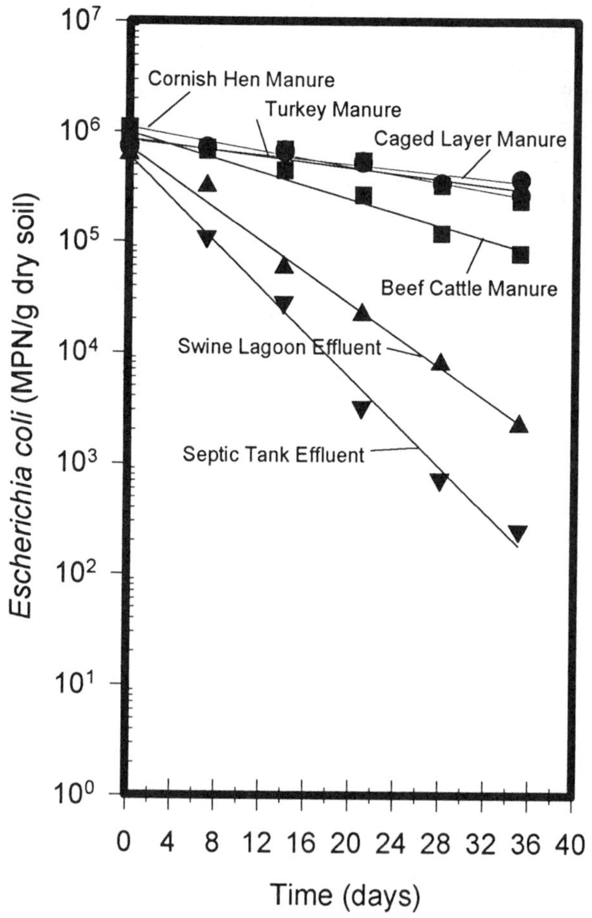

Fig. 1b. *Escherichia coli* die-off in a Captina silt loam amended with poultry manure, beef cattle manure, swine lagoon effluent, or septic tank effluent.

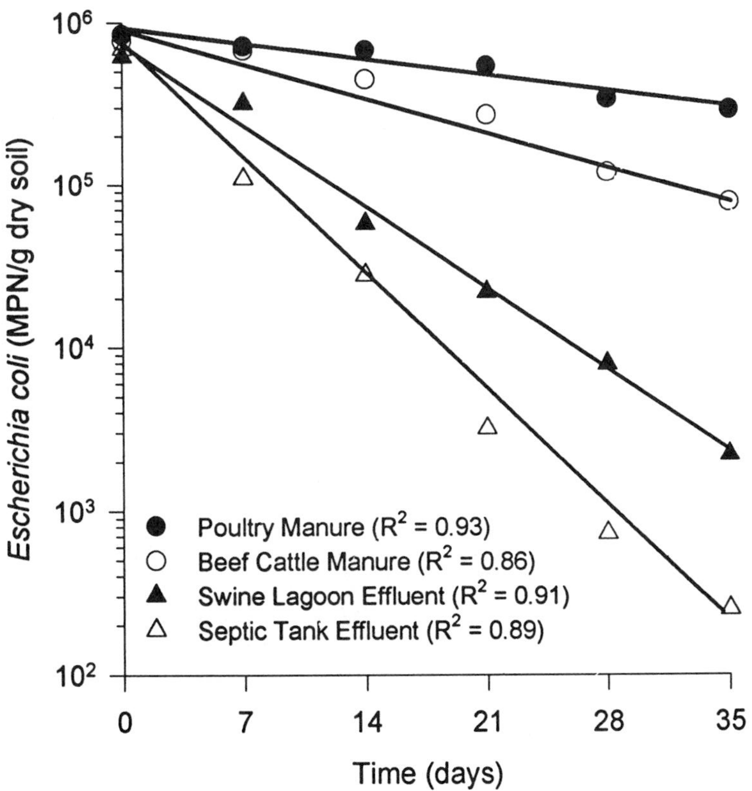

Appendix 9
SAMPLE LETTER REQUESTING COPYRIGHT PERMISSION

The following is a copy of a letter requesting copyright permission for use of the review in Appendix 5 by Dr. Gail Vander Stoep. Although Dr. Vander Stoep has no registered copyright on this review, it still belongs to her, and permission should be requested. Note that it tells her where and how the material will be used and includes a copy of the version that will be published. It also assures her that she will receive credit and acknowledgment for the work.

The letter itself provides a simple form that can confer the official permission once it is signed. If my request were for a number of sections in a work or required lengthy description, I would devise a form to list information on materials requested. Both the *CBE Style Manual* (5th ed., 1983) and the *ACS Style Guide* (1986) provide samples of forms that may be used. See Chapter 12 for complete references to those publications. For the author's files, a duplicate of the request with an original signature by the person making the request is enclosed.

Department of Agronomy
University of Arkansas
Fayetteville, AR 72701
13 May 1996

Dr. Gail Vander Stoep
Michigan State University
PRR Department
131 Natural Resources Building
East Lansing, MI 48824-122

Dear Dr. Vander Stoep:

I am preparing a handbook, entitled *Scientific Papers and Presentations,* that should be published by Academic Press, Inc., San Diego, CA in 1996. The book will be used by graduate students in the sciences and by career scientists. To illustrate points that I make in the text, I plan to use several appendices. I want one of these to be the review you did for the *Journal of Natural Resources and Life Sciences Education* on the manuscript entitled "University-Based Education and Training Programs in Ecotourism or Nature-Based Tourism in the USA."

Can you please grant me permission to reproduce in the book, and in any subsequent editions or reprints thereof, an abridged version of your review. I enclose a copy of the version I want to publish.

With my use of the review, full credit and acknowledgment will be accorded to you. Please confirm your permission for me to use this material by completing the form below and returning this sheet to me. I enclose a duplicate for your files.

Sincerely,

Martha Davis

Permission is granted for use of the materials described in the letter above.

Date ____________________ Signature ____________________

Title ____________________

Appendix 10

IMHOF'S THEORY OF COLORS

The following excerpt is taken from Chapter 4, "The Theory of Colors" (pp. 69–73) in *Cartographic Relief Presentation* by Eduard Imhof. (Permission to use this material has been granted by Walter de Gruyter & Co., Berlin, F. R. Germany.) I would certainly recommend that anyone interested in the use of color consult Imhof's book. Imhof's expertise is in the area of cartography, but what he says about color can apply equally to its use in any visual presentation. In reading the following, you may substitute the term **visual aid** for **map** and think of these principles and rules in regard to making slides, a poster, or other colored displays. Notice the emphasis on the following points:

Subdued colors, neutral colors, or those toned down with gray are best for backgrounds and large areas.

Bright colors and contrasting colors are best reserved for highlighting points of emphasis or for small areas.

Combinations and juxtaposition of colors require that the choices be complementary and harmonious.

Individual choices allow the creator of the map (visual aid) to articulate his or her point of emphasis and make the communication effort unique.

On the harmony of colors and their compositions

The eagle has a keener eye than man, but all that he derives from his perceived image is whether or not it is of interest to him, perhaps edible or dangerous. The human being, although, as a rule, more highly developed in an intellectual-spiritual sense, views the visually perceived world—the forms and colors of things—primarily as a psychological experience. He hangs a colored picture of red roses above his bed. But when the same colors appear in the picture of an open wound, he is horrified.

A color in itself is neither beautiful nor ugly. It exists only in connection with the object or sense to which it belongs and only in interplay with its environment. Concepts of harmony, accord and melody always refer to composition—that is, to the way in which they harmonize or interplay with acoustics or visual elements. With a finely developed artistic sense or as a result of careful education, we can, of course, learn to perceive even non-objective, abstract color compositions as beautiful or ugly. Many simple heraldic figures are, for example, included in such abstract compositions.

Attempts have often been made, and are still being enthusiastically made today, to evaluate drawing and painting by scientific methods. This, however, is very difficult. Here one cannot expect to find laws which can be proven but, rather, demonstrations of generally valid perceptions, experiences, and fashions.

It has been shown repeatedly that great artists have broken through the accepted laws of composition and, in so doing, have achieved extraordinary effects. In the field of painting, intuitive, artistic perception takes over where scientific logic fails.

In spite of these reservations, there follows here an attempt to set down several general *rules of color composition,* insofar as they are significant for maps.

First to be examined are the *combinations of two or more colors,* taken outside the context of pictures or compositions. The latter will follow later.

a) Combinations of two or more colors. Which colors, for example, if presented as adjacent, but separate, similar rectangles or rectangle-like areas, would be most pleasing, and which would clash or be out of harmony? The answers to such questions vary greatly as a rule from one case to another. Fashion, education, psychological inclination and the emotional state at the time are as important here as the existence of any artistic inclination.

In general, the unbroken sequence of colors from the color circle is perceived as harmonious. Compounds of only two colors have harmonious effects if they are complementary colors, that is, if they lie opposite each other in the color circle. The same principle holds for groups of three. Examples are:

Groups of two: Yellow–violet
Yellow orange–blue violet
Orange–blue
Red orange–blue green
Red–green
Red violet–yellow green
Groups of three: Yellow–red–blue
Yellow orange–red violet–blue green
Orange–violet–green
Red orange–blue violet–yellow green

Such duos and triads are even more harmonious if their colors are lightened by white, darkened by black or toned down to a pastel shade by grey, to equal extents. *Subdued colors are more pleasing than pure colors.*

Brown colors are composed of yellow and red, and a small amount of blue. Their harmonious complements are found in colors in which that tone is dominant which is weakest in the brown in question. Examples:

Yellowish brown–blue violet
Reddish brown–blue green

In general, experience has shown that harmony exists between two colors when their subtractive mixtures produce black or grey and, correspondingly, their additive mixtures produce white or grey. This holds true for every pair of complementary colors and, hence, for any two colors whose numerical values in the Hickethier color classification yield, together, a sum consisting of three equal digits.

Leonardo da Vinci said, "There is no effect in nature without cause." Therefore, one searches for explanations of the color harmonies described above. One could find them perhaps in an unconscious striving for order, in the complementary character of colors, or in their contrasts. Perhaps, however, the causes of harmonious effects lie deeper, perhaps in our familiarity with the colors of daylight. We perceive a color composition to be well integrated or harmonious when it results in the white of daylight if mixed additively, or produces grey when mixed subtractively. This applies for the groups listed above. As we have already established, two adjacent colors blend mutually as far as our sense of sight is concerned, and each color tends toward the complementary color of the neighboring hue. *Psychological color perception always tends, therefore, in the direction of composing complementary colors.* This statement appears to be quite significant in explaining the perception of harmony discussed here.

For the same reasons, perhaps, two or more different, bright colors—placed in close proximity to each other—have an unharmonious effect when their combination does not make up white, grey or black.

Examples of this type are the following groups of two:

Red and violet
Violet and blue
Blue and green
Strong yellow and pale whitish red.

One of the most troublesome area colors, for cartographic and other purposes, is *yellow*. It can provide good effects, when it is used in carefully calculated amounts, to give a 'warming' effect to *whole* areas as a base of background tone, but is poor in contrast with white or pale, desaturated bright, yellow-free areas. Yellow and white are similar in that they are the light components of various illumination sources. They stand out poorly when placed adjacent to one another. Midday light and twilight do not appear in a landscape simultaneously. On the other hand, however, yellow goes well with blue, violet and blue-brown probably because of the effects of contrast. Yellow light produces blue or violet shadows.

White, grey and *black*—as *'neutral'* colors, as a surfeit of or lack of daylight—go well with all the bright colors. The compatibility of white and yellow, described above, is the only exception. It is emphasized here once again that the clashing effects of other colors can be subdued by grey, black or dull brown intermediate tones. Unfortunately, however, in maps we can seldom make use of this facility.

So far we have dealt with only the colors themselves and have not gone into *area/size relationships in connection with colors* and into the various color intensities. The harmony of colors can be considerably reduced, or even improved, if their *areas* are *unequal in size* and if the colors are of unequal intensity. *Relationships exist between color intensities and area dimensions.* The purer and richer a color, the smaller its area should be. The duller, the paler, the greyer, the more neutral the color, the larger the area which can be covered. Two bright colors in areas of unequal size go well together only when the smaller area component is strongly colored and the large area is weakly colored. Not only are the qualities of the individual components important, but their quantities as well.

Of a completely different type and based on other phenomena and conventions is a second group of color combinations with harmonious effects. It consists of the *sequence or the change of several continuously graduated colors of one and the same hue* [sic] gradation sequences such as these are brought about by the successive addition of white, grey or black—or even of another bright color. . . . These admixtures give desaturated series, pastel series, shaded series. In nature, they frequently result from differences in distances of observation, or in light intensities, or through atmospheric haze. We perceive them as pleasing or harmonious, perhaps because they have an ordering, grouping, connecting, calming effect and to a large extent reflect the environmental appearance to which we are accustomed.

b) Color compositions. "Tones, harmonies, chords, are not yet music" (Windisch). Only the composition as a whole determines the good and bad of a piece of graphic work. This is also true of a map. Here, of course, one is not completely free to create graphic form. Nevertheless, the cartographer should not blame the chains that bind him for any lack of taste in his work, because he also has sufficient alternatives available to allow his good aesthetic judgment to be employed.

There follow several empirical *rules* which are especially applicable to map design.

First rule: Pure, bright or very strong colors have loud, unbearable effects when they stand unrelieved over large areas adjacent to each other, but extraordinary effects can be achieved when they are used sparingly on or between dull background tones. "Noise is not music. Only a piano allows a crescendo and then a forte, and only on a quiet background can a colorful theme be constructed" (Windisch).

The organization of the earth's surface facilitates graphic solutions of this type in maps. Extremes of any type—highest land zones and deepest sea troughs, temperature maxima and minima, etc.—generally enclose small areas only. If one limits strong, heavy rich and solid colors to the small areas of extremes, then expressive and beautiful colored area patterns occur. If one gives all, especially large areas, glaring, rich colors, the pictures have brilliant, disordered, confusing and unpleasant effects.

Second rule: The placing of light, bright colors mixed with white next to each other usually produces unpleasant results, especially if the colors are used for large areas.

Third rule: Large area background or base-colors do their work most quietly, allowing the smaller, bright areas to stand out most vividly, if the former are muted, greyish or neutral. For this very good reason, *grey* is regarded in painting to be one of the prettiest, most important and most versatile of colors. Strongly muted colors, mixed with grey, provide the best background for the colored theme. This philosophy applies equally to map design.

Fourth rule: If a picture is composed of two or more large, enclosed areas in different colors, then the picture falls apart. Unity will be maintained, however, if the colors of one area are repeatedly intermingled in the other, if the colors are interwoven carpet-fashion throughout the other. The colors of the main theme should be scattered like islands in the background color (see Windisch, 318).

The complex nature of the earth's surface leads to enclosed colored areas, like these, all over maps. They are the islands in the sea, the lakes on continents, they are lowlands, highlands, etc., which often also appear in thematic maps, and provide the desirable amount of disaggregation, interpretation and reiteration within the image.

In this respect, great importance is laid on delineation of areas within maps on the selection of sections, and even on the combination of maps in atlases and also on map legends. Cleverly arranged, legends put life into empty spaces, loosen up uninteresting parts of the image and often produce a balance in the composition.

Fifth rule: The composition should maintain a uniform, basic color mood. The colors of the landscape are unified or harmonized by sunlight.

In many maps and atlases a special green printing color is used for low lying land apart from the blue of the seas. The impression produced is usually poor, the colors of the oceans separating sharply from those of the land areas. In the "Schweizerischer Mittelschulatlas" ("Swiss Secondary School Atlas") and in other maps, however, the low-land green is produced, by over-printing, from the light yellow (used to cover all land areas) and the blue of the oceans. Other mixed tones are also derived from a few basic colors. Only in this way can the unity of light and tone be introduced and disturbing color contrasts avoided.

The idea of a single, uniform basic color mood should not be exaggerated, however. The freshness of colors should not suffer and the contrast between them should not be unduly subdued. The whole map sheet should not appear dull, jaundiced and dead.

Sixth rule: Closely akin to the requirement for a basic color mood, mentioned above, is that of a steady or gradual de-emphasis in colors. A continual softening of area tones is of primary importance in cartographic terrain representation. The natural continuity of the earth's surface demands a similar continuity in its image. Aerial perspective gradation helps this to be attained.

This principle is in no way opposed to a contrary requirement, that of contrast effects. A master reveals himself through the way in which he manipulates the different principles, using moderation on the one hand, but applying deliberate and carefully considered emphasis on the other.

In all questions of form and color composition, one should strive for simple, clear, bold and well-articulated expression. The important or extraordinary should be emphasized, the general and unimportant should be introduced lightly. Uninterrupted, noisy clamor impresses no one. Activity set against a background of subdued calm strength produces a deeply expressive melody.

This is also true in maps. The map is a graphic creation. Even when it is so highly conditioned by scientific purpose, it cannot escape graphic laws. In other fields, art and science may take different pathways. In the realms of cartography, however, they go hand-in-hand. A map will only be evaluated as good in the scientific and didactic sense when it sets forth simply and clearly what its maker wishes to express. A clear map is beautiful as a rule, an unclear map is ugly. Clarity and beauty are closely related concepts.

Appendix 11
EDITORIAL[1]

When you read the following editorial, you may feel that Jay Lehr is being a bit severe in suggesting punishment by stoning for inept speakers. But after you have traveled several thousand miles to attend meetings to enhance your scientific and professional development and have listened to all too many speakers who waste your time with mumbling, reading, showing unreadable slides, or running long minutes overtime (perhaps into your own presentation time), then you may agree with Lehr's pronouncement. At least read the positive suggestions he makes to achieve a successful presentation; he hits upon the main ideas about the audience, the subject matter, the visual aids, and the speaker's delivery. Perhaps his stoning is facetious; his ideas for a good presentation are certainly not.

LET THERE BE STONING!

by Jay H. Lehr

Let there be an end to incredibly boring speakers! They are not sophisticated, erudite scientists speaking above our intellectual capability; they are arrogant, thoughtless individuals who insult our very presence by their lack of concern for our desire to benefit from a meeting which we chose to attend.

We attempt to achieve excellence of written presentation in our journals. We can require no less in our conferences. It is an honor to be accepted as a speaker who will spend the valuable time of hundreds of scientists at a confer-

[1]Reproduced with permission from Marilyn Hoch, Senior Editor of *Ground Water.* This editorial by Jay H. Lehr appeared in *Ground Water* **23,** 162–165.

ence. Failure to spend this time wisely and well, failure to educate, entertain, elucidate, enlighten, and most important of all, failure to maintain attention and interest should be punishable by stoning. There is no excuse for such tedium, so why not exact the ultimate penalty?

Not long ago I became so enraged by a speaker at a conference I moderated, that I publicly humiliated him before 200 hostile attendees. This young man chose to read in a monotone from a secretarial pad, flipping pages for 30 minutes of a scheduled 20-minute speech while complex slides tripped incomprehensibly across a screen behind him. At the conclusion of this group insult, he had the nerve to summarize his presentation, looking up for the first time, by stating that he hoped he had helped us to understand the relationship between the rain in Spain and the crumbling of the Rock of Gibralter, or some such ponderous chain of reasoning. As I awakened the remaining audience, who had not the nerve to walk out as others had, I explained to the young man that he had done no such thing. Trembling as I spoke, I told him and the audience that his paper was an insult which had obviously bored and irritated a kindly group of scientists who deserved better. Those who kept awake refrained from stoning him, though they surely had adequate cause. The young man collapsed into his seat in shock as I proceeded with my vocal condemnation, the audience was pleasantly aghast, and this editorial was born.

What I said then I write now. It needs saying and writing. We have all experienced this insult, and many of us have been guilty of purveying it. It must stop. It is not funny. The penalty must be severe.

I recognized the problem when attending my first conference with my thesis advisor as a graduate student in the '50s. I was appalled at the dreadful presentation I was subjected to. The professor tried to calm my immature ravings by explaining that all meetings were like this and that their value was in the halls, not the auditoriums. I could not see why value could not be attained in both places, but I have remained quiet too long. And so to begin.

The average conference paper is 20 minutes in length. It is not a college lecture where students are to absorb the minutest detail of a subject planned and presented as part of a 10 to 16 week curriculum. Rather, a conference paper offers an up-to-date capsule summary of a particular piece of ongoing or completed research for the purpose of bringing fellow scientists up to date on activities in their field.

A speaker cannot hope to teach the audience the specifics of his work, but he can elicit a valuable appreciation of the research effort and imply the value of the contribution to the growing body of knowledge on the subject. To achieve this he must convey enlightened enthusiasm for his subject and the advances he has attained.

Without exception a presentation with the aforementioned goals can and should be made extemporaneously. A scientist who cannot retain in his head the essence of his latest work can hardly be said to be enraptured by his subject. If a speaker is not excited enough by his area of expertise to weave it comfortably into the fabric of his cognitive thought processes, then how can he hope to excite an audience to an acceptable level of appreciation?

There is never an excuse to read a paper. True, it is the rare speaker who can articulate verbally the same elegant phraseology he commits to paper with the benefit of editing, but fewer still are the preachers of science who can bring the **written word** to life and the audience to the edge of their seats. Better to lower the level of verbal excellence and raise the level of extemporaneous energy. The audience will never know what perfect phraseology they are missing, and the speaker must not allow himself to be frustrated by the inability to turn a perfect phrase in the air. In any case, a paper written for publication never reads well out loud. It's really a different medium. If the speaker excites the audience with his energy, they will want to read the paper later, and then they can rapture in the precision of the written word.

A few notes or an outline are all that are required to maintain order and organization. Slides, of course, can serve the same purpose, but never subject your audience to poor slides just because they serve as an outline for your talk. Poor slides are just a distraction from your hopefully vivid words. They must be brightly lit and convey a simple thought. If you need a pointer to indicate an important concept or location on a slide, it is probably too crowded or difficult to comprehend. If you can't read the print on a slide clearly with the naked eye (reading glasses are permitted) when holding it in your hand, it is inadequate for viewing with a slide projector in any size room with an adequately sized screen.

Never, but never (remember stoning) show a slide and then apologize for it. Don't show it. What did you think of the last speaker you heard say "I apologize for the poor quality of this slide," or "I realize no one past the front row can read this slide," or "I'm sorry you can't read the columns of numbers on this slide but I just wanted to point out . . . "? Point out what, fella—we can't read it, remember? Well, what did you think of these speakers? Dumb at best; *&!*&!*@*, at worst! Resist! Resist! Don't show bad slides! They never help; they always hurt. Don't be afraid to use no slides. Word pictures can be great if you practice painting them with a bit of rehearsal. Many of the best college professors you've heard do just fine in their lectures without slides. You can too—kick the crutch! But if you want to use slides, make them good ones. Good ones are not cheap. You can easily spend a few hundred dollars on a good set of slides for a talk, but look at the dividends:

- Your audience will sit up, take notice, and think you're great, someone special!
- You will invariably find many opportunities to use good slides over and over.

- Your audience paid good money to come and hear you. You probably got in free or at a reduced rate. Reinvest the savings in good slides and give your audience a dividend on their investment.

Don't stay on one slide too long; put blanks between slides if you have a lot to say before the next slide. The old slide is distracting. Don't let the slide lull you into a monotone; keep a high energy level with lots of enthusiasm. When you are giving a paper, you are an actor on a stage. You may be an incredible dullard in real life, putting people to sleep right and left, but at that podium, you're a star. You're an entertainer, an educator; put on a happy face and kick ass . . . or get off that stage. Science is sensational; working in a factory is boring; seeking scientific truth is a turn on, so turn on or you'll turn your audience off. You ought to know, your colleagues have been doing it to you for years. Dare to be different. Use your hands, move around, not to the point of distraction, but look alive! Unless you're a pro and I'm boring you with the obvious, rehearse. Rehearse before a friend or relative and to yourself in the quiet of your mind, on a drive, a run, a swim, a cycle, a daydream, anywhere! Listen to yourself. Your wife, kids, and friends won't want to listen to you; bribe them, they will. If you tell them to be tough on you and let you know what's really bad, they'll love it. Think of the time the audience is collectively giving you. One hundred people times 20 minutes is 33 hours. Don't you owe them a few hours of effort in return?

Get your timing down. No one minds your going a minute or two overtime, but five or eight is inexcusable. Face it, there is extraneous material in your talk. You may love it, but the audience can do without it. Get to the point earlier, and spend more time on the meat and less on the soup and nuts. In the beginning, tell the audience what you're going to tell them. Then tell them, and be sure to leave time at the end to tell them what you told them. It sounds simple, but it works and they will appreciate it.

Make sure you talk into the microphone; tell the audience to let you know if you're too loud or not loud enough. You will lose 20 seconds regaining your composure and properly modulating your voice, but that beats 20 minutes of deafening silence or a rumbling sound system.

Avoid jokes unless you're a stand-up comic. Nothing is colder than a failed attempt at humor. If there is anything humorous in your subject, milk it. That's real and will be well received.

When all is said and done, more is said than done. Don't waste words, but if you must, remember that attitude is 75 percent of nearly everything in life; enthusiasm is at least that in public speaking. Brim with enthusiasm; if you don't have it for your work, how can an audience have it for you? Come alive!

A few words for moderators—you're the master of ceremonies, and you can set the tone for all of the speakers. Show an interest in the session. Open with 30 seconds of well-planned comments. Introduce each speaker with five pertinent points of information which you committed to memory in the past ten minutes, i.e. college degrees and colleges attended, two significant past work affiliations (if pertinent), current work affiliation and activity focus. Do it like you know the speaker well, even if you never laid eyes on him before. You can do it. It takes just three minutes to learn five facts for a short duration. If you're not willing to put in the time, don't accept the job of moderator. When the speaker finishes, keep order during the question period and don't hog the microphone yourself, but do tie it all up with 30 seconds of concluding remarks, if appropriate. A good moderator can really help; a bad one gets in the way, wastes time, and impedes the performance of the speakers he is there to assist.

When on the speakers' platform, unless you have a natural wit and air of showmanship, you cannot afford to be yourself. You must be an actor who is privileged to educate and entertain. The latter must come first or the former has no hope of attainment. But here is one simple rule that can make us surprisingly as comfortable before a group as with a single friend. **Be intimate with your audience.** Make them feel that you are there because you care about informing each and every one of them; no matter if there are 40 or 400, **be intimate.**

When you see the rare speaker who has an air of showmanship that allows him to get into the minds of his audiences, do you comment on how lucky the speaker is to be a natural? Are you sure you can't hope to emulate such a performance? A stage is meant to be acted on, whether to perform in a play or exhort a college student into a broader and deeper understanding of the subject at hand.

We have classrooms in college and stages at conferences because we know that the learning process can be enhanced by animated oral presentations which transcend the capacity to learn from the written page. Unfortunately, most of us achieve less, not more. We deliver an unenthusiastic reading or account which falls more deafly on ears than dead prose falls on our eyes.

Don't get up and do what comes naturally if what comes naturally is a dull, witless, monotonous presentation of unexciting facts. If your work is in fact dull and unexciting, don't burden any audience anywhere with a conference presentation. Publish somewhere if you must, if you can!

If, on the other hand, your work has substance that can be brought to life, do just that. Waste no more time saying you can't do it. Do it or have a colleague do it. There is no longer any excuse to be dull. Regardless of the fact that TV evangelists have given enthusiasm a bad name, **be enthusiastic!**

I studied astronomy under a dullard and thought it was a dead science. Carl Sagan taught me differently. I studied biology under a bore and saw no future in it in my mind. Paul Ehrlich showed me differently. My first economics professor put me to sleep, but Paul Samuelson awakened my interest.

I became a geologist because my earliest mentors, Cary Croneis, John Maxwell, and Harry Hess made the earth live for me. Make your subject—no matter how esoteric—live for your audience if only for 20 minutes.

If everyone takes my message to heart, there need be no more public humiliations and even fewer stonings. But if egotistical, pompous, cavalier, obtuse, inconsiderate ignoramuses insist on ignoring these words to the wise, let there be stoning!

Appendix 12
SAMPLE SLIDE SET

The following slide set was used in a presentation based on the preliminary study described in Appendix 3. Slides were all in color, and the author, Terry Gentry, chose a subdued, dark blue for the background and yellow and white for the text. The photographs and figures add other colors to highlight the presentation. The word slides serve as the outline for the speaker and contain only enough information to reinforce what he will say to the audience. His repeated reference to the literature adds credibility to his own ideas about phytoremediation and microbial diversity in the rhizosphere. Notice how photographs are clearly related to the text and are mixed throughout the set to help convey the message and to provide relief from the word slides. His results are presented entirely with the table and two figures, which he can leave on the screen as he discusses the outcome of his study. These 24 slides are not too many for Terry's 12- to 15-minute presentation. Half this number would be more appropriate if more explanation of complex data were needed.

2

RHIZOSPHERE BACTERIAL DIVERSITY RELATIVE TO PHYTOREMEDIATION OF ORGANIC CONTAMINANTS

Terry J. Gentry

Department of Agronomy
University of Arkansas

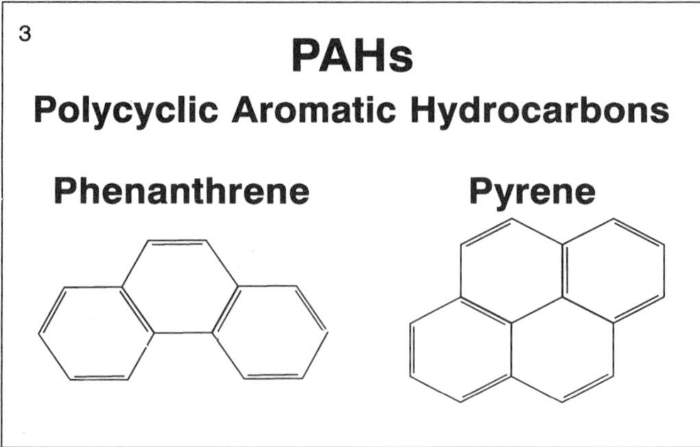

4 **Pathways of Dissipation**

- **Volatilization**
- **Irreversible sorption**
- **Leaching**
- **Accumulation by plants**
- **Biodegradation**

Reilley et al., 1996

6 **Remediation Techniques**

- **Physical containment**
- **Excavation and treatment**
- ***In-situ* treatment**

Lee et al., 1988

7

Factors Controlling *In Situ* Biodegradation

- Soil water
- Oxygen
- Redox
- pH
- Nutrients
- Temperature

Sims et al., 1993

8

Bioremediation

- Use of living organisms to reduce or eliminate hazards resulting from accumulations of toxic chemicals and other hazardous wastes.

9

Phytoremediation

- Use of green plants to remove, contain, or render harmless environmental contaminants.

Cunningham & Berti, 1993

10

Phytoremediation

If pollutants are:

- Near the surface
- Relatively non-leachable
- Not imminent risk to health or environment

Cunningham & Lee, 1995

12

Rhizosphere

- Zone of soil under the direct influence of plant roots and in which there is an increased level of microbial numbers and activity.

Curl & Truelove, 1986

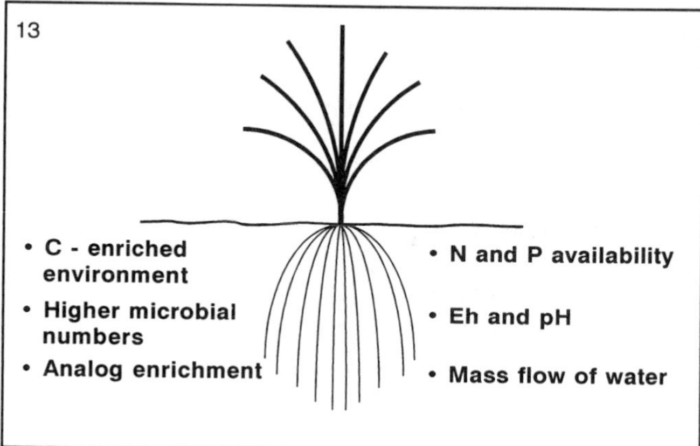

14

- **Increased density of microorganisms**
- **Increased biodegradation**
- **Increased diversity?**

Anderson et al., 1995

15

- **Soil bacteria are the primary degraders of PAHs.**

Shabad & Cohan, 1972

16

Objective

- To assess the impact of the rhizosphere on soil bacterial diversity

17

Materials and Methods

- Captina silt loam and Appling sandy loam
- Bahiagrass and no-plant control
- Growth chamber - 3 wk 16/8 h and 27/16 ± 1°C

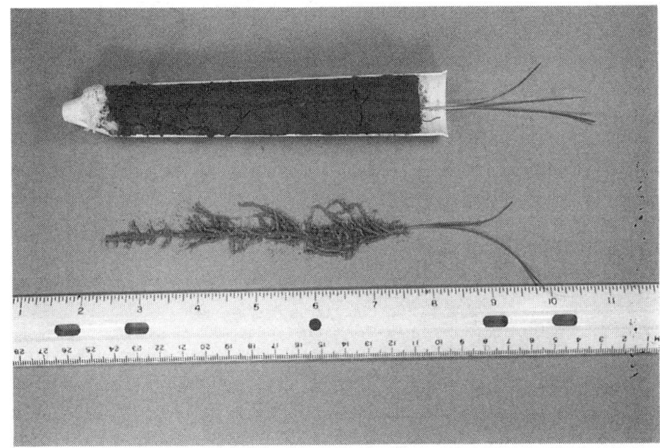

19

- **Total bacterial numbers**
- **200 random isolates**
- **Fatty acid methyl ester analysis (FAME)**

20

Total Bacterial Numbers

Captina silt loam			Appling sandy loam		
Bulk	Rhiz	R/S*	Bulk	Rhiz	R/S
--- 10^6 CFU/g dry soil ---			--- 10^6 CFU/g dry soil ---		
10.0 a**	11.0 a	1.1	5.8 a	18.3 b	3.2

* Ratio of rhizosphere/bulk soil populations.
** For a given soil, numbers with the same lower-case letter are not significantly different at the 5% level.

21

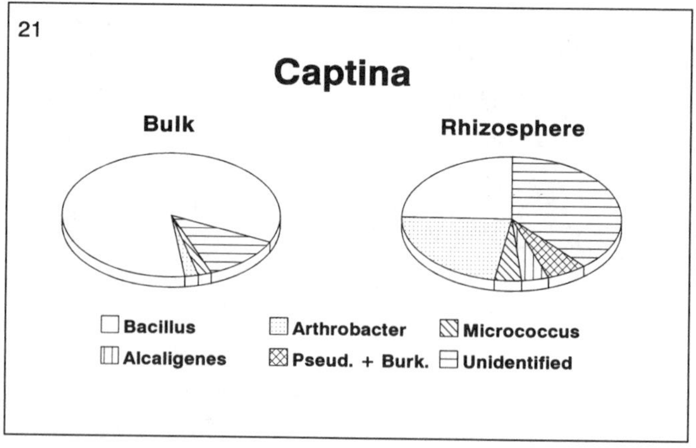

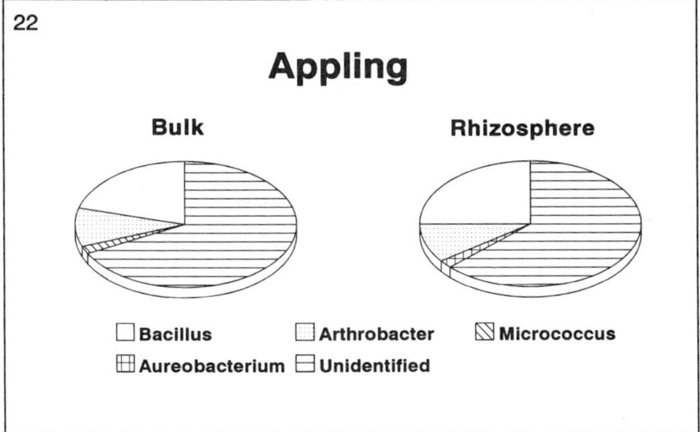

23

Conclusions

- Bahiagrass rhizosphere increased bacterial diversity in Captina silt loam.
- Diversity appeared different between soils.

24

- Increased understanding of the rhizosphere influence on bacterial populations may enhance remediation of PAH-contaminated soils.

Appendix 13
SAMPLE TEXT FOR POSTER

The following poster layout and text are from the poster pictured on page 183. The authors were fortunate to have a large (ca. 4 × 8 ft or 1.2 × 2.4 m) board for display. A smaller board would require that the text be reduced in a manner similar to that of reducing an abstract (see Appendix 7). Probably with a board half this size, the introduction would be no more than one sentence of justification preceding the objectives. Methods and results would have to be cut despite their importance. Reference to the literature would probably be omitted or very limited, and discussion might be no more than a sentence or two.

Notice the layout of this poster (Fig. A13-1). Sections are balanced with the

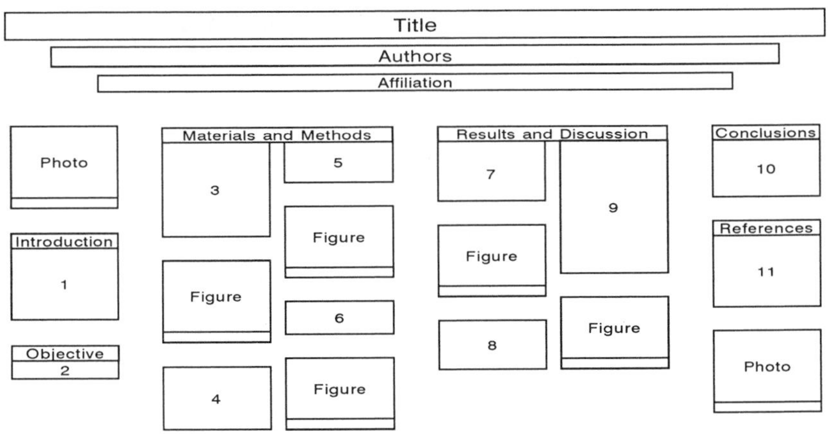

FIGURE A13-1
Layout for poster in the photograph on p. 181.

first and last columns as single rows. The center portions also balance in the same way with double rows connected by their headings. The fact that the columns are not the same length and each piece of the poster is not the same size adds aesthetic interest that is not present in a monotonous repetition of the same sizes and levels throughout. Note also that spacing within sections communicates; spaces are wider between sections, but both columns of the materials and methods and the results and discussion sections are grouped closer together.

The text is presented below with numbers corresponding to those on Fig. A13-1. It is in short segments with only Section 9 being somewhat long, and much of both the methods and results is communicated via figures. Clear statements of objectives and conclusions are set apart with their own headings to make it easy for the reader to know what was planned and the basic outcome even without reading methods and results. Photographs help to describe the site and the system, but they also have aesthetic value and serve as relief from additional text that might be needed to describe the site and system.

Notice that the central idea, as expressed in the objectives and conclusions, is carried into detail with photographs, figures, and text to make this poster a reader-friendly display.

TEXT FOR FIGURE A13-1

RENOVATION OF ON-SITE DOMESTIC WASTEWATER IN A POORLY DRAINED SOIL

D. C. Wolf, E. M. Rutledge, M. A. Gross, and K. E. Earlywine

University of Arkansas and Arkansas Department of Health

1. INTRODUCTION

Justification

Approximately 29% of the U.S. population disposes of domestic wastewater through on-site disposal systems (Miller, 1984). During the past 10 years, the rate of construction of soil absorption systems for on-site wastewater disposal has increased by 20% (Reed *et al.*, 1989). However, septic systems are frequently cited as sources of water contamination.

Introduction to topics in the poster

To protect water quality, it is essential that the soil renovate the domestic wastewater before it replenishes water supplies (Reneau *et al.*, 1989). In poorly drained soils, hydraulic limitations can result in septic systems that fail to adequately purify the effluent and, thus, impose an adverse impact on water quality.

2. OBJECTIVE

Specific plan

The objective of the study was to use a low pressure distribution system with tile drains to evaluate the renovation of septic effluent in a poorly drained soil.

3. MATERIALS AND METHODS

Methods are important and unique to this study. Therefore, they are described in detail with 3 figures to illustrate.

A clear statement of methods provides credibility for the study.

Septic effluent was pumped from the dose tank into filter field absorption beds indicated as B1, B2, B3, and B4 in Fig. 1. The beds were loaded at a rate of 18 L/m^2/day and received the septic tank effluent through 0.48-cm orifices in 3.8-cm nominal diameter, schedule 40 PVC pipe. The system distributed effluent evenly over the beds by maintaining approximately 60 cm of head. Tile drain trenches shown as T1, T2, T3, T4, and T5 in Fig. 1 were located beside and between the absorption beds. A cross section of the system is shown in Fig. 2. The drain trenches and the absorption beds were separated by 100 cm of undisturbed soil. Each sand-filled tile trench contained a nominal 5-cm diameter Hancor "Turflow" slotted drain pipe located 10 cm from the bottom of the trench. The tile drains discharged into a sump where each tile was sampled. In addition to the filter field system, background tile drain trenches denoted as T6, T7, T8, T9, and T10 were sampled.

Description of the site

4.

The study site had a 0% slope and was located on a Calloway silt loam (fine-silty, mixed, thermic Glossaquic Fragiudalf). Poor drainage and a high seasonal water table were indicated by mottling throughout the soil profile which had a fragipan or Btx horizon at 116 cm. The amounts of precipitation added to the soil during the study are given in Fig. 3.

Methods of sample collection and analysis

5.

Samples were collected from the dose tank, tile drains in the filter field, and background tile drains biweekly from November through May in 1987–88 and 1988–89. Samples were filtered through a 0.22-μm membrane filter and analyzed with standard methods for concentrations of NH_4–N, NO_3–N, Cl^-, and total soluble organic carbon (TOC). Electrical conductivity was also determined. We determined fecal coliform numbers on unfiltered samples using the membrane filter procedure.

Statistical methods

6.

Statistical analysis included analysis of variance and mean separation to compare chemical and biological parameters in the filter field tile drains (T2, T3, and T4) and background tile drains (T7, T8, and T9).

7. RESULTS AND DISCUSSION

Specific results and comparison to the literature add credibility to the study.

The septic tank effluent had mean TOC, NH_4–N, NO_3–N, and Cl^- concentrations of 55, 41, <1, and 50 mg/L, respectively, and an electrical conductivity of 0.84 dS/m. The values are typical for septic tank effluent (Cogger and Carlile, 1984; Piluk and Hao, 1989).

Specific results with reference to the figure

Results indicated that 1 m of soil was highly effective in reducing TOC levels in the wastewater effluent. Samples collected from the tile drains in the filter field absorption bed showed a 96% reduction in TOC (Fig. 4). Background tile drains had significantly lower TOC levels than tile drains in the filter field, and values were not significantly different between years.

→

Notice that results are separated here by a figure that is referred to both above and below it.

8.

The Cl^- concentrations decreased from 50 mg/L in the septic effluent to 35 mg/L in the tile drains from the absorption bed to 8 mg/L in the tile drains from the background area (Fig. 4). The electrical conductivity values showed a similar trend.

The NH_4–N did not move through the soil as indicated by levels of 1 and 0 mg/L in the drain tiles from the absorption bed and background areas, respectively (Fig. 4). Walker *et al.* (1973) reported that NH_4–N levels decreased to low levels within 20 cm of the filter field beds in four soils studied.

More specific results, but the figure expresses the details.

9.

The mean NO_3–N level in the tile drains from the absorption bed was 4 mg/L and significantly greater than the 1 mg/L in the background drains (Fig. 4). Mean NO_3–N levels in the tile drains showed a significant increase from 1 mg/L in 1987–88 to 5 mg/L in 1988–89 while the Cl^- levels were not different between years.

General assessment of results with support from the literature

Limited discussion

The combination of NH_4–N retention by the soil exchange sites and denitrification of the NO_3–N resulted in a reduction of the NH_4+NO_3–N/Cl^- ratio from 0.8 in the septic effluent to 0.1 in tile drain samples from the absorption beds. The reduced ratio would indicate that denitrification was occurring. Stewart and Reneau (1988) studied a low pressure distribution system in a soil with a seasonally fluctuating water table and reported active denitrification. It appears that nitrification occurs when the soil is aerobic, and denitrification takes place when the soil becomes saturated (Reneau, 1979).

More specific results with reference to Figure 5 and comparison with the literature

Soil renovation of fecal coliforms was also demonstrated with geometric mean values in the tile drains in the absorption beds and background tile drains of 18 and 3/100 mL, respectively (Fig. 5). Reneau (1978) used tile drains in wet soils to study fecal coliform movement from septic systems and reported that bacterial numbers decreased with distance from the septic system, and the decrease was described by a logarithmic equation.

10. CONCLUSIONS

General statement of outcome

Analyses of water samples collected from the septic system dose tank, tile drains in the filter field absorption beds, and background tile drains showed that, in a poorly drained soil, a low pressure distribution system with tile drains resulted in a significant renovation of domestic wastewater within 1 m

Suggestion for further study

of soil. Future research will involve quantitating water volume in the tile drains to enable us to assess the NO_3–N loading rate (volume × concentration) of the septic system.

11. REFERENCES

Complete documentation for citations in the text

(Type size for these citations was somewhat smaller than for other text.)

Cogger, C. G., and Carlile, B. L. (1984). Field performance of conventional and alternative septic systems in wet soils. *J. Environ. Qual.* **13,** 137–142.

Miller, D. W. (1984). Sources of ground-water pollution. In *Protecting Ground Water, the Hidden Resource,* pp. 17–19. U.S. Environmental Protection Agency, Washington, DC.

Piluk, R. J., and Hao, O. J. (1989). Evaluation of on-site waste disposal system for nitrogen reduction. *J. Environ. Eng. Div. Am. Soc. Civ. Eng.* **115,** 725–740.

Reed, B. E., Matsumoto, M. R., Wake, A., Iwamoto, H., and Takeda, F. (1989). Improvements in soil absorption trench design. *J. Environ. Eng. Div. Am. Soc. Civ. Eng.* **115,** 853–857.

Reneau, R. B., Jr. (1978). Influence of artificial drainage on penetration of coliform bacteria from septic tank effluents into wet tile drained soils. *J. Environ. Qual.* **7,** 23–30.

Reneau, R. B., Jr. (1979). Changes in concentrations of selected chemical pollutants in wet, tile-drained soil systems as influenced by disposal of septic tank effluents. *J. Environ. Qual.* **8,** 189–196.

Reneau, R. B., Jr., Hagedorn, C., and Degen, M. J. (1989). Fate and transport of biological and inorganic contaminants from on-site disposal of domestic wastewater. *J. Environ. Qual.* **18,** 135–144.

Stewart, L. W., and Reneau, R. B., Jr. (1988). Shallowly placed, low pressure distribution system to treat domestic wastewater in soils with fluctuating high water tables. *J. Environ. Qual.* **17,** 499–504.

Walker, W. G., Bouma, J., Keeney, D. R., and Magdoff, F. R. (1973). Nitrogen transformations during subsurface disposal of septic tank effluent in sands: I. Soil transformations. *J. Environ. Qual.* **2,** 475–480.

Captions for figures

Fig. 1. Design of the low pressure distribution septic system with tile drains and the background tile drains used in the study.

Fig. 2. Cross section of an absorption bed and tile drain trenches.

Fig. 3. Monthly precipitation amounts during the 2-yr study.

Fig. 4. Mean chemical concentrations in the septic effluent, tile drains from the filter field, and background tile drains.

Fig. 5. Geometric mean fecal coliform numbers in tile drains from the filter field and for 1987–89.

Captions for photographs

Sand-filled tile drains were installed between the absorption beds and in a background area.

A low pressure distribution system was installed in a Calloway soil.

Annotated Bibliography of Select References

Booth, V. (1993). "Communicating in Science: Writing a Scientific Paper and Speaking at Scientific Meetings," 2nd ed. Cambridge Univ. Press, Cambridge, UK.

Booth's publication is a succinct book dedicated to T. W. Fline (Those Whose First Language Is Not English). One section is on writing scientific papers and another on speaking at professional meetings. His discussion on use of the language includes important conventions in grammar, punctuation, and other mechanics. He describes manners, style, visual aids, and delivery techniques that are important to a speaker's rapport with an audience. In other short chapters, he discusses use and misuse of numbers, troublesome aspects of the language for those for whom English is a second language, preparing figures and copy for print, and notes on the differences in British and American English. His 78 pages contain a great deal of information for both the writer and the speaker in science.

Committee on the Conduct of Science (1989). "On Being a Scientist." National Academy of Sciences, National Academy Press,Washington, DC.

This booklet is an effort by the National Academy of Sciences to recognize the human element—human judgments, human values, human error, human relationships—in scientific research. The work is directed toward students beginning their scientific careers.

> Much of the first half of the booklet looks at several examples of the choices that scientists make in their work as individuals. . . . The second half deals largely with questions that arise during the interactions among scientists . . . A final section touches upon the social context. (p. 1)

The booklet views issues that the scientist faces relative to values in the largely unwritten code of professional conduct and personal integrity in science. It provides some detail on questions concerning scientific values, human decisions, treatment of data, sources of error, fraud, plagiarism, appropriate credit, obligations to society, and other issues that every scientist can encounter.

Council of Biology Editors (CBE) Scientific Illustration Committee (1988). "Illustrating Science: Standards for Publication." CBE, Bethesda, MD.

This well-illustrated book presents perspectives on preparing and publishing artwork for science. "Our goal is not to teach the fine points of executing and illustration; rather it is to guide the judgment of an illustration's worth" (p. vii). In guiding our judgment, the book does teach us a great deal about the fine points in communicating with illustrations. Subject matter includes information on drawings and photographs of biological specimens; designing and drawing graphs and maps; and type styles and sizes, use of tone and color, halftones, line art, and other details in illustration processes. It also provides information on materials to use in the production of illustrations, getting an illustration ready to print, and legal and ethical considerations involved.

Council of Biology Editors (CBE) Style Manual Committee (1983). "CBE Style Manual," 5th ed. CBE, Bethesda, MD.

This manual presents conventions of technical style for the sciences based on standards established by both "international and U.S. organizations concerned with science or with information services" (p. xvii). It provides information on planning and writing the manuscript for publication, prose style (grammatical conventions), and style in documentation and tables and figures. It offers guidelines for dealing with proofs, indexing, copyrighting, and ethical conduct in authoring and publishing a paper. It includes stylistic conventions used in special fields of plant sciences, microbiology, animal sciences, chemistry and biochemistry, and geography and geology.

Council of Biology Editors (CBE) Style Manual Committee (1994). "Scientific Style and Format: The CBE Manual for Authors, Editors, and Publishers," 6th ed. Cambridge Univ. Press, Cambridge, UK.

More comprehensive relative to stylistic details than the 5th edition, this manual contains conventions for all of the scientific disciplines. This reference is a good attempt to simplify publication style as much as possible and to provide a uniform style for all the sciences. It may be the most comprehensive scientific style manual available relative to the conventions such as punctuation, abbreviations, capitalization, symbolization, and references. It provides information on general stylistic conventions as well as specific matters in science such as cells, species, and systems where the conventions for style are unique to the subject. The manual provides information on format in journals and books including form for tables, figures, and indexes. It does not, however, contain any instructions for planning, writing, and submitting papers for scientific publication such as are included in the 5th edition.

Day, R. A. (1994). "How to Write and Publish a Scientific Paper," 4th ed. Oryx Press, Phoenix, AZ.

A "cookbook" on preparing manuscripts for publication in scientific journals, this book deals with all the elements that constitute the paper from the title to the references. It contains information on other issues in the publication process, such as submission and review of the paper, and short chapters on other kinds of scientific communication such as review papers, posters, and oral presentations. Day presents pleasant, even entertaining, discussion on the use and misuse of English and avoiding jargon; some information on what constitutes scientific writing; ethical and legal issues; and helpful appendices on matters of publication style.

Hodges, E. R. S. (1989). "The Guild Handbook of Scientific Illustration." Van Nostrand–Reinhold, New York.

"This book is designed to be a reference for the scientific illustrator, other artists, and scientists who do their own drawings or hire illustrators" (p. xii). The emphasis here is on both technique and subject matter. Details in the use of pencil, line and ink, airbush, carbon dust, watercolors, and other methods are presented. Subject matter illustrated includes both the animal and the plant kingdoms with physical depiction of birds, mammals, fossils, medical subjects, and other objects illustrated in science. Hodges describes processes for illustrating involving microscopes, charts, maps, graphs, and diagrams. The book concludes with a section on the business of science illustration including information on copyright, contracts, and other business matters.

Imhof, E. (1982). "Cartographic Relief Presentations." de Gruyter, New York.

> First and foremost, this book is directed to cartographers, the actual designers of maps; but it should also interest topographers, geodesists, surveyors, photogrammetric engineers, geographers and scientists as well as teachers of cartography and geography and, finally, anyone with a love of good maps. (p. v)

Imhof provides guidelines to ways of rendering terrain to make maps more reliable and readable. He deals with concepts of how to show topographical relief with skeletal lines, contour lines, shading, hachuring, and color. Chapter 4, "The Theory of Colors," is valuable for anyone using color in publications or presentations. The book contains very good examples.

Macrina, F. L. (1995). "Scientific Integrity: An Introductory Text with Cases." ASM Press, Washington, DC.

Macrina's text provides a background for discussing issues on integrity in science. For the graduate advisor, the chapter on mentoring is excellent, and for students and other researchers the information on record keeping can be very helpful. Questions of integrity regarding authorship and peer review, conflicts of interest, the use of animals or humans in experimentation, genetic technology, and the ownership of data with patents and copyrights all are presented well by Macrina and other authors of individual chapters. This text would be good for a class on ethics in science.

O'Connor, M. (1991). "Writing Successfully in Science." Harper Collins *Academic*, London.

Most of this book is centered around the writing for publication in the scientific journal. O'Connor contends that "Journal articles constructed in this formal way have become the basic units of research publication and are the model for many other kinds of writing in science" (p. ix). She covers planning, writing, preparing data in tables and figures, revising, submitting, and checking proofs of such an article. She also includes information on preparing and making presentations with slides and posters. The book provides information on writing grant proposals, theses, and review articles and is good for writers whose first language is not English.

Reynolds, L., and Simmonds, D. (1983). "Presentation of Data in Science." Nijhoff, The Hague.

"The aim of this book is to provide the amateur with an awareness of basic principles and practices in the preparation of visual materials" (p. xxii). The book contains basic design principles relative to such matters as legibility of type (size, style, and spacing); illustrations for publications, slides, and posters; tables and figures; overhead transparencies, tape/slide programs, and television; and materials and equipment for artwork. It provides little on the use of computer technology, but basic principles remain the same.

Smith, R. V. (1984). "Graduate Research: A Guide for Students in the Sciences." ISI Press, Philadelphia.

Smith's book "is designed for self instruction" (p. ix). His text serves to orient students to ideas about graduate research programs by discussing the student's commitment to the program, the principles and ethics involved in scientific research, and planning and time management. He deals with practical matters of library research, writing, preparing theses, presenting and publishing papers, getting grant support, and even getting a job.

Stock, M. (1985). "A Practical Guide to Graduate Research." McGraw–Hill, New York.

A helpful handbook for the graduate student in science. Deals with the "realities of how research is actually conducted" and carries the student "through the graduate program and through development, execution, and completion of the research project" (p. vii). This completion requires oral and written communication. Stock presents a practical, realistic view of the graduate research program with advice from choosing a topic and an advisor to the job interview seminar. She includes ideas on writing grant proposals, theses, and journal articles and on presenting talks and visuals.

Tichy, H. J. (1988). "Effective Writing for Engineers, Managers, Scientists," 2nd ed. Wiley, New York.

Tichy's handy reference book for all kinds of writers provides a thorough discussion about organization and getting started writing. The book contains numerous examples to illustrate the ideas presented. It has a section on grammar, punctuation, and diction. Also, it discusses the literary prose style or character of writing that results from sentence structure, word choice, the arrangement of ideas, and other elements of construction. It is a valuable guide for organization and clarity in any kind of writing. It contains a good appendix on troublesome words in English.

Zinsser, W. (1988). "On Writing Well: An Informal Guide to Writing Nonfiction," 3rd ed. Harper & Row, New York.

Zinsser writes about the principles involved with any kind of writing, and the book itself serves as an outstanding example of writing well. These principles lead to "using the English language in a way that will achieve the greatest strength and the least clutter" (p. 6). His principles of readability take into consideration the audience, the diction, construction and grammatical usage. Then he discusses forms used to write with unity about people and places and about subjects such as science, business, or sports. He provides evidence that good writing has universal qualities not to be diluted by the discipline in which it communicates.

Index

J

L

M

N

O

P

Q

R

S

V

W